BRITAIN
ACCORDING TO
KALEB

Kaleb
Cooper

QUERCUS

First published in Great Britain in 2023 by
Quercus Editions Ltd
Carmelite House
50 Victoria Embankment
London EC4Y 0DZ
An Hachette UK company

QUERCUS

A CIP catalogue record for this book is available
from the British Library

HB ISBN 978 1 52942 481 2
Ebook ISBN 978 1 52942 483 6

10 9 8 7 6 5 4 3 2 1

Designed and typeset by Julyan Bayes at Us-Now Design & Art Direction
Illustrations by Julyan Bayes
Printed and bound in Italy by LEGO S.p.A.

Papers used by Quercus are from well-managed forests and other responsible sources.

CONTENTS

Dedicated to my beautiful daughter
Willa Grace Cooper

FOREWORD

Hi, it's me again. I'm back, which is a happy surprise. I did a book, about how I see the world, and so many people liked it and bought it that the publishers asked me to do another one. I was glad to say yes, because the world is all well and good, but there's so much more to talk about.

'Wait, do you mean space?' they said. I explained, no. At least, not this time. Maybe the next one. That would be pretty cool. This time, I wanted to look at my own country. I wanted to look at all the weird and wonderful traditions and events – the really old ones, the brand new ones – that happen all around Britain. The things that tell us who we are, where we come from, what we're like, and what we like.

There's an Italian guy called Alberto Grandi who said, 'A tradition is nothing but an innovation that was once successful.' That's exactly what I think. Whether we've been doing something for eight hundred years or eight, the fact that we're actually doing it is the important bit.

Britain has so much variety, imagination and straight-up fun in its customs, contests and amusements that I couldn't think of anything more enjoyable than to go through them region by region and see what I make of them. So here they are. Thanks for joining me!

Chapter One

Oxfordshire & Surrounds

CHEESE ROLLING

Let's face it, when people think of Gloucestershire, they tend to think of one thing. A pack of lunatics with a death wish chasing a cheese down a hill. That is quite unfair, and not at all representative of the county. The great majority of the Gloucestershire people are not lunatics who chase an eight-pound round of double Gloucester down hills. In fact, they are lunatics who quite fancy chasing a cheese down a hill but just haven't got around to trying it yet.

Living your best life, one fractured limb at a time.

Now, I'm from Oxfordshire. North-West Oxfordshire, right near the Gloucestershire border, which means Gloucestershire is a full five minutes away by car, making it a completely different world. Until recently, wild horses couldn't even have dragged me out of Oxfordshire. Not that we have any wild horses, which is one of the things I like about it. We've got enough to deal with – sheep and whatnot – without wild horses getting in the way. And I'm sure they feel the same in Gloucestershire. Wild horses would probably just join in and kick the you-know-what out of you when you're trying to run down a hill after some cheese, and who wants that?

The last thing you need. Also, the last thing you'll ever see.

But then I somehow became famous – although it still boggles my mind that that could even happen...

So I was invited to leave Oxfordshire and go to the cheese-rolling contest at Cooper's Hill. The first thing I noticed is that the farmer whose field is at the foot of the hill is coining it at a fiver a car for parking, which of course I respect. The second thing is it's not actually a hill. It's a f***ing cliff. A vertical lawn.

At the top of the 'hill', it's chaos. At the bottom, it's indescribable carnage. The first race isn't so bad, but they're just getting warmed up. The second is for kids, and they go uphill, for safety – then when they get to the top they all come straight back down again, for fun, the mad little demons. By the third race, for women, someone had already snapped their

'Woo-hoo! Winners' podium, here I come!'

ankle, just by watching – they got hit by the cheese, which was travelling at light speed. I even got injured myself, when a stone came down the hill and cut my leg open. The winner knocked herself out, then started cheering along with everyone else when she came round.

In the fourth race, one guy landed head-first next to me, stoved all his teeth in, then got up and asked me for a selfie. I had to tell him, 'Mate, you haven't got a face right now.'

When you look at all the insane things people get up to in the countryside, perhaps it's just something to do with being from the countryside in the first place. Even so, I'm not sure there's anything quite as mad anywhere else. I find myself wondering how it all started. Maybe we just really, really like cheese. Still, there are limits. I mean, don't get me wrong, I like a bit of cheese. Let me put that another way. I like a lot of cheese. Tell the truth, I *love* a lot of cheese. But I wouldn't kill myself to get it. I might kill you. And your friends. And everyone you know. Let's face it, if you stood between me and cheese, I'd probably do away with the lot of you, without even blinking – and that, I think we can all agree, is a perfectly normal relationship with cheese. But I wouldn't kill *me* to get it, for the sole reason that I would then be unable to

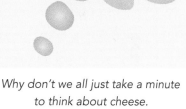

Why don't we all just take a minute to think about cheese. There, doesn't that feel better?

eat the cheese. The lovely, delicious, delectable cheese … sorry, where were we? Right!

I have the cheese! I'm sorry so many of you had to die for this, but, you know, them's the breaks. Literally, for most of you.'

So I completely understand the motivation. Nowadays, you get people travelling from all over to get their hands on that cheese. One man came from Toronto in Canada and won the men's event. I can only assume they don't have any Double Gloucester where he's from, which is a heartbreaking thought. Those poor people.

I reckon the whole thing began because the cheese comes in a wheel, and wheels roll downhill. I mean, of course they do, right? I know everyone goes on about how clever the person who invented the wheel must have been, and fair enough – where would we be without them? Not lying in a heap at the bottom of Cooper's Hill with two broken legs, I suppose, but you have to take the rough with the smooth. All I'm saying is, let's wait until somebody invents a wheel that rolls uphill and

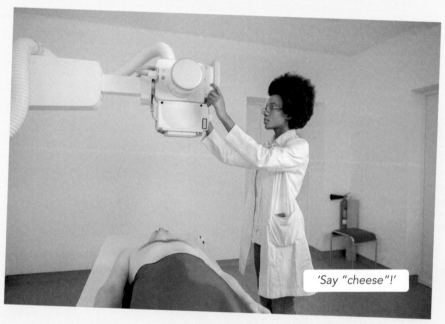

'Say "cheese"!'

Now you're talking.

then we'll see who the greatest genius of all time is. Anyway, my theory is, one day somebody was carrying a load of cheese over the top of the hill, and a wheel of cheese dropped off the pile and down it went, and then some people went after it trying to catch it, and the rest is history. OK, strictly speaking, the rest is *Casualty*.

It's a decent-sized wheel they send down there nowadays, but I can't help thinking the risk is still a little excessive in proportion to the reward. What you really want is a tractor wheel of cheese. And come to think of it, a tractor to ride on when you're going after it. Although, when don't you need a tractor?

THE MORETON-IN-MARSH SHOW

Not actually the Moreton-in-Marsh show, but hey, close enough.

This is one of the biggest one-day farming shows in England, and it's really local to me. I even have my own stand now. I started in 2022. If you want to be anybody, you literally have to take a stand. Because everyone else does. You're like a farming version of Spartacus.

You really have to look the part, so everybody gets dressed up. The only time you might ever dress better is your wedding day, and even then, I wouldn't count on it. A really nice checked shirt, your good jeans, a gilet and your workboots, the steel toe-cap ones. Not your actual workboots, which are covered in so much mud you can hardly see them. Your other pair, the ones you've got as a backup, in case something happens to your first pair, like they get sucked down into a bog or something.

You want it, we got it

DOG RING
LIVESTOCK
POULTRY
COUNTRYSIDE
FOOD HALL
CTIONS
HORSE
TOILETS

A lot of people come to this show – and they do it for three main reasons. Number one: to have a cracking day. Number two: to meet people they don't see all year. Everyone will be

'Then we'll be staying home, thank you very much.'

there, somewhere. It's just a matter of finding out where. It's like a rabbit warren. Only without the shagging, obviously. It's not that kind of event.

The third reason is to show animals. That's a huge deal; it has a massive following. They'll take their cows, their sheep, their pigs. I always personally try to avoid the sheep, but normally I get dragged in there some way or another. Things you can't avoid: death, taxes and sheep. The cows are amazing. I love the cattle side of it. People take showing animals very, very seriously – they don't get paid for it, but it's much more than commercial anyway; it's a

Suave.

matter of pride in their work. They've been rearing them all year. Or much longer: rearing cattle takes a long time. If you're a beef farmer, you've got to wait two years before you get your money. If you're a breeder, you've got to wait two years before the animal has a baby, and then another year to grow the baby, and so on.

And then there are the medals. All those little kids going out there, showing their animals – it's a real achievement for them. They want to show that their animals are the best in the world; you cannot put a price tag on their passion. I would spend all day watching the cattle walk around the ring if I could, seeing the smiles on the people's faces – it's utterly amazing.

'More like, udderly amazing. I'll be here all day folks.'

Then there's all the time taken behind the scenes, the training, the washing. If you've ever seen any Highland cattle, well, you can probably imagine what happens if you wash them. It's like me with a perm, as anyone who read my first book (still available at all good bookshops and websites, and apparently a few dodgy ones as well) will remember. You have to soak the hair, then you give it a bit of a blow dry, and *bam*, it goes all frizzy. Trying to control that is hard work. Anyone who's got long curly hair will understand that.

It's a bit like a muscle car meet. Especially with the bulls, everybody's parading around trying to show they've got the biggest ... let's just say, doors or wheels. Keep it nice. And it's the same for me, because the reason why I can't spend all day watching the cattle, which I'd like to do, is that I take

'My dear, I just can't do a thing with it.'

my agricultural machinery, my tractors, and line them all up, show them off, really – so everybody can see what I can do, and to gain some business. But again, I also want to demonstrate that my tractor is better and cooler than your tractor. Because it is.

All my friends know the low rider ...

TETBURY WOOL SACK RACES

This goes back to when Tetbury was a big deal in the wool trade. Being sheep-related, it wouldn't usually be my thing. But the wool's not part of the sheep any more, so I'm happy to talk about it. The contestants race up a steep hill carrying big sacks of wool – sixty pounds for men, thirty five for women. Personally I wouldn't want to be carrying another sixty pounds of anything. I work on a farm and I carry hay bales all day, every day already. That said, I don't mind watching if somebody else wants to do it. However, I'd rather go and feed some cows, as at least I know I'd get paid at the end of the year. The wool industry is very low at the moment and it's costing farmers more to shear the sheep than they get back for the wool. But you have to shear them anyway because if you don't, they'll suffer in the hot summer months, and might even get

maggots. So I guess this is probably the best thing to use the wool for right now, honouring the area's history.

"Join Blue Peter", they said. "You'll have a brilliant time", they said.

BIBURY DUCK RACE

I've been to this, and it's really good fun. I'll definitely take the little ones when they're old enough. On Boxing Day each year, they have two races where they put a load of ducks into the Coln river. One race has one hundred and fifty realistic-looking decoy ducks. You pay a tenner and if your duck wins, you get to decide which charity gets all the ticket money. Or you can pay fifty pence for one of two thousand plastic ducks in the second race. Normally my duck goes, 'Oh, look at that shiny thing over there,' and heads off in the wrong direction, which I never knew an artificial duck could do, but mine have a talent for it. There's a fascination to watching anything you throw in a river, like a twig or a leaf, so a plastic duck is twenty times more fascinating. There you are, yelling at a fake duck to go faster in the water. Come to think of it, I probably look like an idiot. But at least I'm among all the other idiots. We're all there together, shouting encouragement at plastic ducks. It's great.

'Come on, number 1,723!'

SUMMER SOLSTICE AT STONEHENGE

For me, the longest day of the year is mainly a way to get more work done. It's a real celebration of countryside, just as it is for the people who come to this. They get to see spectacular views of the sunrise, which any farmer can tell you is always a wonderful thing. Mind you, I'm not so sure about going and dancing around a load of stones that come up out of the

ground. I've ploughed plenty of stones. The Cotswolds is on brash, which means the ground just under the soil is mostly rocks. If I was farming stones, I'd be minted. And if every time I ploughed a stone up out of the ground, I started dancing around it naked, I'd never get anything done. Every couple of minutes, it seems like you hit one. That's a lot of stones, and a lot of naked dancing – and let's be honest, nobody wants to see that. I'd probably end up on YouTube. And it would probably be one of my mates who put me there.

Probably best not to ask.

Chapter Two

South-West England

South-West England is very special to me. Mainly because I've actually been there. And it takes a lot to get me out of the Cotswolds. I have to go to London from time to time, now that I'm a globe-trotting international celebrity. By 'globe-trotting', I mean the other bits of the Cotswolds. By 'international', I mean places that aren't the Cotswolds. And by 'celebrity', I mean bloke who gets recognized by other country people because we're almost never on the telly, so it's easy to stand out.

Basically it's this bloke, and me, and he gets all the babes.

Obviously, London doesn't suit me. Too many buildings, not enough fields. But I'll say one thing for the place: it's Party Central. And when I went to Cornwall, I found it was the same thing. All right, Party South-West, I suppose. Which is the best of both worlds. The farmers there are like inner London party animals in the countryside. Honestly, I'm not sure how they ever get any work

Needs more flat caps and gilets.

done. Nothing seems to happen in the winter months except for partying. I went in November and it was awesome. They love it.

Another thing I like about the South-West is the accent. It's the most country accent there is. It's like a country accent on steroids. I've got a country accent myself, obviously, but I tip my hat to this one – it truly is the classic country accent, and you've got to love it for that.

Ain't no party like a West Country party.

Another funny thing is that the classic farmer voice and the stereotypical pirate voice are thought to be exactly the same. If that were true, in the South-West, every day would be Talk Like A Pirate Day. It seems to be the law that all pirates have to come from Cornwall or Somerset. I had no idea that piracy was such a well-regulated field. Obviously, I'm all in favour of well-regulated fields. If I had anything to do with it, no field would be anything else. But I'd always thought the point of being a pirate was that, unlike being a farmer, you really didn't have to worry about red tape. No entry requirements, apart from being bloodthirsty and rocking an eyepatch and parrot. But no – apparently, if you can't do a convincing 'Arrrr!', you're out. Seems harsh to me, not to mention discriminatory, that you can't be a Scouse pirate, for example, but I don't make the rules.

'Come 'ed, soft la!'

I was really impressed by the way South–West farmers use every scrap of land, too. My theory is that it's because they've got so much coastline. I grant you, everybody's got more coastline than where I'm from. Round Chipping Norton way, we only need to look at a pond to feel seasick. But in the South-West, the sea is everywhere. I think they value the land more there because the sea is always getting at it. There isn't a square inch that they don't grow stuff on. I was so impressed when I visited for a week's holiday, I ended up buying ten pounds of veg. 'We'll never eat all that in a week!' said my missus. 'Don't you worry, love,' I told her. And I had a fair go, but I admit we ended up taking some back home, which is a bit like coals-to-Newcastle really.

'How are you supposed to resist when they put it in a horn? A HORN, I tell you!'

WASSAILING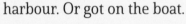

The biggest form of partying they have, the one that's really popular in the South-West region, is called wassailing. When I first heard about it, I thought they said 'sailing'. Which sort of made sense, although I was surprised they did it for fun, while drinking. If I was drunk in charge of a boat, they'd have to summon the RNLI to rescue me before I'd even left the harbour. Or got on the boat.

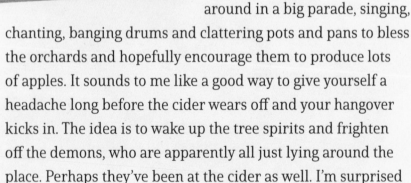

It's all going swimmingly.

Wassailing is something quite different. It starts with you drinking what we farmers refer to, in technical jargon, as a sh*t-tonne of cider. In fact, almost everything they do in the South-West starts that way. Then, you put on the daftest clothes you can find and dance around in a big parade, singing, chanting, banging drums and clattering pots and pans to bless the orchards and hopefully encourage them to produce lots of apples. It sounds to me like a good way to give yourself a headache long before the cider wears off and your hangover kicks in. The idea is to wake up the tree spirits and frighten off the demons, who are apparently all just lying around the place. Perhaps they've been at the cider as well. I'm surprised they don't frighten off the trees and all. If I was a tree, I'd probably be so terrified I'd refuse to do anything – no blossoms, no apples, nothing. But it seems to work for them, so their trees must be made of harder stuff than me. I suppose, being trees, they would be.

To paraphrase the Duke of Wellington, we don't know what effect they will have upon the demons, but by God, they frighten us.

You've heard the phrase 'a piss-up in a brewery', but the wassailers are clever: they cut out the brewery – or in this case, the cidery – and go straight to the source. I think I've worked out how this all started. Basically, the South-West has a cider-based economy. They've got all these apples that they have to do something with. So, they make them into cider. Then they have to come up with an excuse for people to drink loads of cider, so what did they think of? An event that results in producing more apples! It's brilliant, a perfect self-sustaining system – and you get to have a massive bash while you're at it. My flat cap is off to them.

It's a tough old job, but somebody's got to do it.

They only do it once a year, around Twelfth Night, which is impressive in itself, because the state of me after Christmas, I couldn't even look sideways at a cider, let alone drink a wassail bowl of it. But they just go right at it, and meanwhile the demons are probably saying, 'Sod this, we're off to the pub,' where – you guessed it – they'll be drinking even more cider. I know city people usually think people in the South-West are some kind of ultra-yokels, but they couldn't be more wrong. They're as sharp as you like and they've got everyone playing their game – tourists, demons, pirates, the lot.

'Just a half for me, thanks, I've got work in the morning.'

THE HUNTING OF THE EARL OF RONE

As far as I can make out, this is quite similar to wassailing, only they must have laced the cider with LSD. Each May, in a village in Devon called Combe Martin, one person dresses up as the earl, although if you went into the House of Lords looking like that, you'd be gently but firmly escorted to the exit and placed in the care of highly qualified professionals. (And that place has a very high tolerance for eccentricity, so that tells you something.) Then, the rest of the village spends the whole weekend looking for him. Or rather, I should say 'looking' for him, because they never find him until Monday, which is rather convenient. 'Any sign of him yet? No? Oh dear. Let's have another cider while we have a good long think about where he might possibly be.'

In fairness, we wouldn't be in too much of a hurry to find him, either.

Then, when they do find him, they have a big parade – I'm spotting a theme here – which is full of things that make the actual earl look as if he's just some feller who popped out to the corner shop in his slippers. There's something that's a cross between a giant birthday cake and Mr Blobby and, honestly, if I came face to face with that, I'm not sure I'd ever sleep again.

No wonder the earl was running away. We would have.

After that, they 'shoot' the earl, then bring him back to life, over and over again, until finally at sunset they chuck him in the sea. It's all fun and games now, but I bet it started off as something quite serious. Then cider got involved, and that was probably for the best. It may look like a living nightmare, but at least nobody actually croaks it.

WORLD STINGING NETTLE-EATING CHAMPIONSHIP

This takes place the Saturday before the summer solstice, in a place called Marshwood in Dorset. I've heard Dorset is known as 'Thomas Hardy country', which I thought must be great, because I love those old black-and-white comedies with him and the other one, the skinny guy.

'Why, yes, I did enter this contest of my own free will. Why do you ask?'

It turns out Thomas Hardy was a man who wrote books about people going through awful suffering in the countryside. So a nettle-eating contest sounds right up his cart track. I couldn't think of anything worse. To this day, if I get stung, I still get a dock leaf and rub it on myself. I don't think it helps, but running around like a headless chicken looking for a dock leaf is a handy distraction from the pain. A bit of mind control. You think you're doing something about it, so it's fine. When the truth is, it's not fine. It's bloody agony. And that's just from brushing against one little nettle leaf. So, actually eating them? It makes your tongue swell up and blacken, and I don't see how that's good for anyone.

'Well, this was tremendous fun, and now, if you'll excuse me, I have to die.'

It's held at a place called The Bottle Inn, and so it should be, because you'd need a lot of bottle to eat even one of the f***ers, let alone around a hundred feet's worth of them, which is what it takes to win. And which is a hundred times more than it would take to kill me.

If you want to know the difference between the South-West and my part of the country, then you can think about the fact that they have a nettle-eating competition while we have an onion-eating competition. I love onions, don't get me wrong, but as soon as I cut one up, I'm in floods of tears, thinking about when my tractor blew up on me. I can see how an onion-eating contest might be quite therapeutic: get your emotions going, go on, clear 'em out. So, I'm in. And I'd rather cry than die.

You can't hide your tears behind dark glasses, you know.

COMBE DOWN TUNNEL ULTRA

When I found out about this one, that's when I knew that people in the South-West really are different. It's an old railway tunnel near Bath, that's the longest foot and bicycle tunnel in the country, over a mile and a half from one end to the other. Anyone else would just go through it as quick as they can and carry on with their day. Not this lot – they spend two days down there, in pitch darkness, running back and forth a hundred times. You can't have anyone to help you in any way. You can't run alongside anyone else (although how on earth they could tell if you did, I have no idea). You can't wear headphones. One contestant went temporarily blind and they thought that was great. I can only guess that they looked at the nettle-eating championship and thought, wow, that's horrible, gruelling and pointless – what can we do that's even worse?

We would have shown you what it looks like in pitch black but . . . you know.

DORSET KNOB-THROWING CONTEST

This is a family book, and my first thought was, I am not going to touch this one. So to speak. I was relieved to find out that a Dorset knob is actually a savoury biscuit. In a place called Cattistock, they have people who can chuck them over seventy feet, underarm, while keeping one foot on the ground. It must be great training for cricket. I still think it's a waste of biscuits, though. Just think if we could get them together with the cheese from the cheese-rolling round my way. So much tasty knob cheese to get down your gulle. Hang on. Let's walk that thought right back to where it came from.

'Get away from me, you knob!'

THE GRAND NATIONAL SHEEPSTAKES

This is a race for sheep.

It is held in Bideford, Devon.

I am never going to Bideford, Devon.

This must be the sheeplechase event.

Chapter Three

South-East England

When you think of South-East England, or at least when I do, I think, well, it's the bits around London, isn't it? Which is why I've tended to avoid it. Although I've been to agricultural shows in Buckinghamshire. Country people there love to point out that they come from Buckinghamshire. That's fair enough. It's nice there, and it's a nice word. It trips off the tongue. No wonder they like saying it as much as they do. Also, I think they want you to know they've got nothing to do with London, that they're just as rural as you are. They've got sideburns, and a really broad accent. Same with the men.

'Buckinghamshire bugg – handlebars. Buckinghamshire handlebars, we calls 'em.'

What the South-East wants you to think it's like...

I like the shows they have. I'd go more often, but they're in the middle of July and I'm busy with the harvest. There's always a field full of trading stands, and little girls and boys having pony races. But don't be fooled. The South-East wants you to think it's all very wholesome. But when you look at some of the stuff they get up to, they're pretty hardcore.

...what the South-East is really like.

I think there might even be parts that are feral, and I say that as someone who lives near Gloucester, which has things in it even David Attenborough wouldn't believe. So, credit to them. They haven't let London stop them being as mad as country folk everywhere. In fact, even London itself has some mad stuff. They've got the Hackney Clown Service – an annual church service attended by clowns, which sounds absolutely bloody terrifying. Also, every year the Freemen of the City of London drive their sheep across London Bridge. When I first heard that, I admit, I got the wrong end of the stick.

'Is this strictly necessary?'

Then again, so do the poor sheep, getting herded around London. Plus, being sheep, they'd probably go straight over the side of the bridge and into the Thames. It takes a lot to make me feel sorry for sheep, but that'll do it. Just this one time, I'm on their side. Although I'm not sure which side I'd be on in the Oxford vs Cambridge Goat Race, which was until recently held on the same afternoon as the boat race, at Spitalfields City Farm. I bet you'd have been able to tell which were the country goats and which were the London goats. The London goats would all be wearing white trainers. Which I suppose might be considered an unfair advantage.

The one in the pumped-up kicks is already out of sight.

WIFE-CARRYING RACE

This is in a place in Surrey called Dorking. I've never been out that way, no surprise – too close to London. If anyone said to me, ever so politely, 'Would you like to go to a carry-your-wife competition?', I'd say, just as politely, 'No, thank you.' There's really nothing else to be said. But what I'd be thinking is, 'ARE YOU BLOODY MAD?' My missus would hate it as well. It's a disaster waiting to happen. If you go and ask your wife, 'Do you want me to carry you upside down for four hundred metres up and down a steep slope over a load of hay bale hurdles?', it'd be almost as dangerous as actually doing it. I doubt she'd be quite as polite as me. Both sides of that conversation would make it a hard 'No!' I can't even run a hundred metres carrying nothing. I'd be dead from that alone.

Whatever these dudes want to carry off, we're not about to argue.

I've had a look at the event website, and this is what it lists by way of risks: 'Slipped disc, broken legs and arms, limb dislocations, neck and spinal damage, facial injury, skull fracture, hernias, and other sundry injuries and illnesses, potentially including death.' You've got to admire their approach – talk

'Comfy, darling?'

about truth in advertising! 'It could kill you, it probably will, what do you say?' I suppose there are people who look at that and think, 'Only "potentially"? I like those odds!'

I'll say this for it, though: it's inclusive. There's no rule that says the 'wife' has to be married to you, or female. So, well done there. And for all the dire warnings, they do take safety seriously. Helmets are compulsory. For the wives. Apparently the whole thing was inspired by Viking raids in the eighth century, when the terrifying Norsemen used to carry off local wenches (I bet it was the blokes who had the helmets then, and not the lasses). When you look at some of the competitors, that makes sense.

INTERNATIONAL BIRDMAN

I'm not gonna lie, I love this. A load of people attempting to fly off a pier in Bognor Regis or in Worthing. Unfortunately, gravity really doesn't work in our favour, does it? I can go down there and tell them that, if they want me to, but they obviously don't, because they've got a whole event dedicated to proving Isaac Newton wrong.

'But what if it fell upwards tho'?

I like their ambition. I like that you spend all year working on a flying suit or contraption, then in two minutes you've fallen to the bottom of the sea. Then again, there's a £30,000 prize for reaching a hundred metres. When I heard about that, my first thought was, I'm all for it, I'll be there next week. It's amazing what you can decide is a good idea when there's a bit of dosh involved. Money really does talk: it can even talk you off the end of a pier and into the English Channel.

There are three classes of entry. I'd class all of them as bonkers,

'Ker-ching!'

myself, but that's what's fun about it. In ascending order of daftness: there's the Condor class, which is essentially people on a hang glider, so they can genuinely fly to

Born to fly or die trying. OK, just to die trying.

some degree. Then, there's the Leonardo da Vinci class, which is for people who've built machines at home, to try and do self-powered flight. And finally there's my favourite class, which is the Kingfisher class, which is basically just people in fancy dress jumping into the sea. So, you can either go all out for the thirty grand on a glider. Or you can try to scoop the jackpot on some mad device you've invented. Or you can just go, f*** it, I'm gonna hurt myself while looking as flamboyant as possible.

It's all for Instagram, I bet – although why they did it before Instagram was invented, who knows? Maybe that's why they had to invent Instagram in the first place, because otherwise it was all going to waste. Plus, they raise money for charity, so I like that. Let's be honest, though, dressing up as a bird and jumping into the sea is its own reward. There's a purity of purpose there. Even better is dressing up as something that can't fly. Like a chicken, or a hedgehog – you know, the least airborne thing ever. Recognizing the pointlessness of it all, but going ahead anyway. I can only applaud that. From a safe distance, of course.

MALDON MUD RACE

Now here's a fancy dress event that's much more in my line: running through the mud, in the Blackwater Estuary. True, I can't actually run, in normal circumstances – unless the cows get out, or something like that. But these aren't normal circumstances. These are muddy circumstances, and that suits me down to the sodden, shoe-swallowing ground. I would absolutely give this a go. In mud, I'm like a pig. It's almost like my natural habitat. I'm in the pig pen so much, I can guarantee I know how to get across that mud ASAP. I know it looks like some kind of muddy hell on earth from the outside, but that's only to people who don't know how mud works. For a mud demon like me, it'd be a doddle.

I'm not sure what I'd wear, though. Maybe I'd go as a scarecrow. On second thoughts, no. I'd go as Captain Jack Sparrow. It'd be like that famous scene of him being chased along the beach, only completely covered in mud. Which, let's face it, is probably what it would really be like if it ever happened. It would reflect the difference between what Maldon is famous for, which is posh gourmet sea salt, and what really goes on there, which is lunatics wallowing around in all the world's mud.

What Walt Disney shows you...

...and what Walt Disney doesn't.

HERNHILL WHEELIE BIN RACE

They call Kent the Garden of England, so I suppose they must have loads of wheelie bins to put all the garden waste in. This event involves drinking beer and... well, that explains it, really. Apparently, it all started at the pub. As do so many things that should probably just have stayed at the pub rather than be brought out into the world. But fair play to the regulars at The Three Horseshoes in Hernhill. They came up with the idea of pushing each other in a wheelie bin, then they turned it into a competitive sport. It's just a matter of time before they get accepted into the Olympics, and turn pro. I'll probably have to buy my lad whatever brand of trainers sponsor the World Wheelie Bin championship.

'Out the road, nippers – elite athletes coming through!'

Yes, it's a tough old game.

I hate taking the wheelie bin out. When you're on a farm, you've got to walk all the way down the farm track to put your wheelie bin where it needs to be. It's not fun. But on the other hand, I've had lots of practice, so I reckon I'd have a good chance in this. Bring it on. Plus, I like the way they decorate the wheelie bins. The whole thing has a *Last of the Summer Wine* vibe to it. On second thoughts, this might be another one that's better as a spectator sport. I'd love to watch it, but I'm not sure I fancy getting clobbered around inside a pimped-out wheelie bin. Leave it to the pros, I say.

WEIGHING THE MAYOR

This makes more sense than you might think. Nowadays, it's normal to be on, let's say, the husky side. That's what happens when you get comfortable. Look at me. I'm engaged. I've got kids. I only need to glance sideways at a pork pie and the waistband button flies off my jeans and stuns a nearby sheep that has somehow discovered a whole new way to destroy itself. But in the old days, the only larger people were the king, the gentry and corrupt local officials. So, if you wanted to know if your mayor had been scoffing your taxes rather than sorting out the potholes in the village cart tracks, there was a simple way to find out. You weighed them. And in those days you could probably bribe somebody with a pork pie. The whole national GDP was measured in scratchings. If somebody had their hand in the till, it would come out greasy.

If I'm honest, you can probably still bribe me with a pork pie. I'd best not ever run for office in High Wycombe, because never mind raiding the public purse, I'm more liable to raid the public fridge. By the time they got me on those giant scales, the reading would be the Olde English equivalent of 'No coach parties, please'.

'Good news! He's lighter than the last one, so we won't have to raise your council tax.'

Chapter Four

Wales

The sea: large, wet, dangerous and surrounded by rocks.

I know two main things about Wales. One is that it's beautiful. Gorgeous scenery. Rolling hills everywhere. Rugged coastlines.

Although, being from the Cotswolds, any coastline looks rugged to me, even if it's covered in little kids and sandcastles. I avoid coastlines as much as possible. They've got the sea there. You might fall in.

The Welsh coastline is so scenic that even a guy called Fred, from the football club I support, Manchester United, likes to hang out there whenever he can. And he's from Brazil, where they have the most famous beach in the world, so I'm guessing he knows a thing or two about the seaside. I like that a footballer from Brazil is known as Fred. It's as if I called myself Kalebinho João Gonçalves de Porto Ferreira every time I went in goal at a charity match. Fred says he likes going to Llandudno every time he gets the chance. Perhaps because it's the place in Wales with the fewest Ls in its name, so it's easier to remember.

The other thing I know about Wales is that it's full of sheep. Three sheep for every one human being. Which, given that Wales's population density is around one-third of England's,

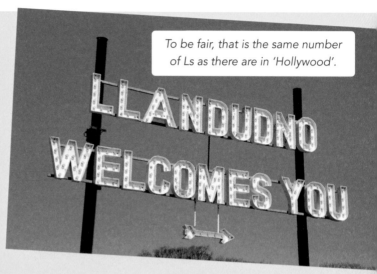

To be fair, that is the same number of Ls as there are in 'Hollywood'.

means that you've got exactly the same chance of encountering a sheep in Wales as you have a human being in England. Right there, that's a strong incentive for me not to go there.

'Yeah, that's right, pal, how d'you like them odds?'

Come and have a go if you think you're hard enough.

That said, there are more castles per square mile in Wales than anywhere else in Britain– which is definitely something I like about it. I'd love to go to a festival in a castle. That sounds like lots of fun.

But it does make me wonder why they built so many castles in the first place. Was everybody always trying to invade Wales? And if so, why? Not to put Wales down or anything, but they've mainly got hills and sheep and that doesn't sound like a good enough reason to raise an army and go and fight a bunch of people who are tough enough to want to live in the hills, with the sheep. Which, trust me, is no picnic. Maybe it's just that the Welsh really like castles, and they can't see a hill without thinking, 'You know what that needs? Let's stick a castle on it. Now that's lovely.'

All the clichés, stereotypes and insulting jokes that non-country people aim at farmers in general, farmers in general aim at Welsh farmers in particular. There's a saying, 'Wellies and sheep', which I am not going to explain in a family book. In fact, I wouldn't explain it in any book. But even though we all roll our eyes whenever we hear that stuff, we then go and say the same things ourselves about farmers in Wales. I'm not proud of that, so, sorry, Wales.

It's just that there are so many sheep there. Almost all the straw I produce goes to Wales for the livestock. And because all the place names are unpronounceable, every lorry driver I talk to, when I ask them where they're going, they say, 'Mate, I have no idea.' Honestly, I'm not even sure the Welsh know how to pronounce them. I think they just make it up as they go along, then laugh at you when you've gone away still trying to find it. They probably think it serves us right for all the stupid jokes, and I suppose it does.

You're just winding us up now, Wales.

WORLD BOG-SNORKELLING CHAMPIONSHIPS

I like the way they say 'World'. As if there's loads of other bog-snorkelling championships taking place across the globe, but this is the one that really matters. That appeals to the entrepreneur in me: top-drawer marketing, that. 'Oh, no, you don't want to go the bog-snorkelling championships in Paraguay. Absolutely not up to scratch.'

This takes place in Llanwrtyd Wells. Apparently, it started as a pub bet, which is a definite pattern for a lot of these events. What it does tell you is that people take a pub bet seriously – and so they should. If they say they're going to do something, they're true to their word. I was in the pub the other day playing darts, and I bet somebody a pig. He won, so I gave him a pig. Then he gave me some work, so it was good for everyone.

The point is, nobody ended up having to doggy-paddle sixty yards and back through a trench four foot wide and fifty foot deep in a peat bog. That's

If you win a pig in a bet, is that bringing home the bacon?

doggy-paddle, because proper swimming strokes aren't allowed. Oh no. Doing the butterfly through an actual bog would be far too easy. This whole thing makes me think that maybe Welsh people looked

'I've snorkelled bogs in Switzerland, Indonesia, Alaska and the Congo, and let me tell you, this is the real deal.'

at that mud race in Maldon and thought, 'Nah, not nearly difficult or messy enough, how can we go one better? I know – instead of wading through muck, let's swim through it.' I think I might just about be able to manage it. But only as long as I was allowed windscreen wipers on my goggles. Otherwise, no chance. I'm not sure even my extensive pig pen experience would see me through this one. Peat bogs are not really my thing. Obviously, there are people who think differently, and good luck to them. I'll just stand over here and watch. 'Over here' being Chipping Norton.

'What, and miss all this?'

MARI LWYD

Why?

This doesn't mean people in Wales are big fans of Victorian music hall. OK, they might be, but that's not what this is about. Oh no. No such luck. If a lady in an old-fashioned frock showed up at your door around Christmastime and started singing 'My Old Man (Said Follow The Van)', you might think she was an eccentric carol singer, but you'd still give her a mince pie and a glass of mulled wine or whatever.

No, what happens here is that THE SINGLE SCARIEST THING I HAVE EVER SEEN IN MY ENTIRE LIFE COMES ROUND YOUR HOUSE AND – FORGIVE MY LANGUAGE – BUT, F***ING HELL.

MARIE LLOYD.

She seems nice enough.

Lots of places in Britain have a kind of hobby horse tradition. And they're usually OK. Somebody dresses up as a horse and goes cantering about the place. It's all good-natured fun – the kids love it. Only in Wales would they put an actual horse's skull on a pole and give it to somebody wrapped in a white sheet so they can go door to door and, I don't know, TERRIFY

WHY?

THE LIVING SH*T OUT OF YOU. I'd be dead. That's all there is to it. 'Oh, there's the doorbell. It must be the neighbours come to wish me Merry Christmas and a Happy New Year, I'll take them a cup of tea.' Open the door, see that thing on the doorstep – thud. That's me gone. Stretched out on the hallway carpet. Dead as the doornail that would be the second-last thing I'd ever see. The last being THIS UNSPEAKABLE ABOMINATION FROM THE DEEPEST CIRCLE OF HELLFIRE COME TO DEVOUR MY VERY SOUL.

Oh, sweet baby Jesus, now it's brought all its friends! Why, God, WHYYYYYY?

It gets worse. Mari Lwyd means 'grey mare', which reminds me of an old folk song that goes, 'The old grey mare she ain't what she used to be', and I'll say she isn't. For one thing, SHE USED TO BE ALIVE AND NOT HORRIFYING. This is a literal nightmare: it's night, there's a mare – if that isn't where the word comes from, it ought to be. The Mari Lwyd starts singing at you, when it gets to your house, which frankly is all you bloody need, and you have to engage it in a 'pwnco' contest, which also requires you to make up witty rhymes to outdo the horse, despite you at this point being exactly as dead as the horse is, but not so articulate. So, basically, you have to have a rap battle with a dead horse in a bedsheet. And this all happens in a place called Llangynwd, which is what you get when you ask an AI generator to come up with the most Welsh name ever. If I had any doubt at all that Wales is winding us up by way of revenge for all the uncouth jokes, this has settled it. Well played, Wales. Well played.

UNDERGROUND TRAMPOLINING & MINI-GOLF

They've got a lot of old coalmines in Wales, which again is very different from where I live. Here, if we dig a hole, it's usually so we can put something into it, not take anything out of it. Also, it's not half a mile deep, because there's nothing we want to get rid of that badly.

I've heard they have raves down some of the mines, which sounds brilliant. Although seeing as I like to dress up for a party, I'm not sure how well that would go. You put on your lovely checked shirt and blue jeans and best trainers and go down a coalmine and, well, you might as well be a goth. So the Bounce Below trampoline park at Llechwedd Caverns sounds a much better bet. First, because they used to mine slate there not coal, but second because, trampolines! Underground! Boingggg! You just have to hope the trampolines aren't too good, because you don't usually have to worry about hitting your head on the underside of the earth's surface. I'm worried I might shoot up through the ground looking like a naked mole rat.

Didn't we see this in one of those Matrix movies?

I'd probably get the world's steepest cable car to the underground mini-golf course instead. I love a bit of mini-golf. And I like the entrepreneurial spirit of the whole thing even more. You know – 'We've got all these giant holes in the ground we're not using, what are we going to do with them?' – then thinking of a money-spinner to put down there. I really admire that. Count me in, definitely.

'I am absolutely fine with this situation and enjoying myself immensely.'

MAN VS HORSE MOUNTAIN RACE

Well, what do you know – here's another event that started after a pub bet in Llanwrtyd Wells. I'm going to go ahead and assume that that's basically what the entire local economy is based on: sitting in the pub thinking of mad stuff to do that people will come to see. And it seems to be working for them, so I'm not going to argue the toss. This race pits humans against horses with riders to see who is faster over mountain terrain. It starts in the town centre, and goes up into the Cambrian Mountains for twenty-two miles. It's been going for over forty years, and in all that time a human runner has won it three

To really make it interesting, they should also have a man carrying a horse.

times, which is three more than I would have expected, and also three more times than I'm ever going to try it, either on foot or on a horse. Still, these horses are alive, though. You've got to give them that. It may be mental, but at least it isn't terrifying.

WORLD BATHTUBBING CHAMPIONSHIPS

Guess where this one started. Go on. Take a wild guess.

That's right. In Llanwrtyd Wells. In the pub.

I'm no expert on bathtubs, that's for sure, but the one thing I do know is that the water's supposed to be on the inside, not the outside. I might have to pop down to Llanwrtyd Wells and give them a few pointers. Trouble is, I'd only end up in the pub and next thing I know I'd have invented a bizarre new sport over a pint or ten and I'd be the World Self-Catapulting Champion or something, last seen in orbit around Pluto.

When you're having as much fun as this guy, you get to laugh at him. But not until then.

CROSSWELL RAS BECA

This is interesting because it's a race that commemorates the Rebecca Riots of the nineteenth century, when poor farm workers disguised themselves as women in order to protest against their living conditions and unfair taxes. So now they have an annual run in women's clothes to mark the event. Which essentially means that Wales invented RuPaul's Drag Race two centuries early. That's forward thinking for you.

'Honey, I don't know who Rebecca is, but I'm always a riot.'

CAERPHILLY BIG CHEESE & GREAT CHEESE RACE

Oh my God. You know how I said I'd like to go to a festival in a castle? Well, it turns out there is one, and it's all about cheese. This is going to have to be the greatest thing ever invented. Whatever unflattering things I've said about Wales, I take it all back. Have I mentioned that I love cheese? Because I really, really love cheese. I would love to go to this and nick all the freebies. I'd put on two stone in a day and come out looking like a block of cheese, and I would not care. Plus, instead of chasing cheese down a hill like they do round here in Gloucestershire, they're much more sensible here and they have a race where they carry it on truckles, so it doesn't get ruined and you can still eat it. I just hope they carry it Caerphilly. I said, I just hope they carry it Caerphilly. No? Oh, suit yourself then.

It's a metaphor for life: the faster you run the race, the sooner you eat the cheese.

Chapter Five

The Midlands

I used to think that I lived in the Midlands, but apparently it starts just north of me. Which explains why I don't miss out words all the time. I was talking to someone from the Midlands the other day – we were sorting out a cow – and he said, 'Put cow in crush.' I said, 'Do you mean, "Put the cow in the crush,"?' The further north you go, the more words they leave out, until you get to Yorkshire and it's just random syllables accompanied by a hard stare.

In case you're wondering what a 'crush' is, it's a kind of cage to hold cows in so they can't move around too much. It keeps them safe – quite literally stock-still – when they're being checked out or treated by the vet, for instance. It keeps you safe, as well, seeing as you don't want to end up underneath a tonne of live cow. I know it sounds like a car compactor for cattle, but don't worry. It's not a device to turn a cow into an Oxo cube in one easy manoeuvre.

Cattle crushes were evidently more space-efficient in the old days.

Or you could just bring these and do a lucky dip.

So, basically, everything directly west of me is Wales, where I have decided I am never going to go because of the horse skull horror (see page 62), and everything directly north and north-east is the Midlands, where I need to bring my own spare words with me to add to other people's sentences. I'm ordering two gross of 'the' to last the weekend before I even step across the Oxfordshire border. But I definitely want to go, because they may be short of words, but they have some great events.

THE WORLD HEN-RACING CHAMPIONSHIPS

This is brilliant, I absolutely love it. Count me in. There are loads of events and loads of traditions where the main point seems to be to maim or kill yourself, but this is completely different. It's held in a village called Bonsall, in Derbyshire. There's a twenty-metre track and the hens have to cover it inside four minutes in the heats in order to qualify for the race. Although let's not use that word – chickens don't like anything that reminds them of an oven. You stand at the finish line and try to encourage them with tasty treats, like corn and meal worms – or maybe frogs and mice. I've seen a chicken

eat a mouse, and it was brutal. Not sure why that doesn't just encourage all the contestants, as well as your hen, but really, who cares? The only thing that could make this better is if you put little T. Rex arms on all of them. Partly because I always like to remind everyone that they're little T. Rexes in the first place, and partly just because it would look hilarious.

Yes, this is how you improve the unimprovable.

If I were taking part, I'd get properly into it. I'd train my hens beforehand. Put go-faster stripes on them, like I have on my tractor. I'd take them for runs every morning. They'd be doing their laps, I'd be standing there in my tracksuit with a stopwatch. Or maybe I'd sit on my tractor with a megaphone. It'd be like *Chariots of Fire*, only for chickens.

It's good, it's classy, but it needs more poultry.

77

The best bit of all is anyone can go along without a hen and hire one for the day. Me being me, I see a business opportunity there. I'll bring along thirty highly trained chickens and rent them out. If your entrant wins, in addition to the rental fee, the bag of grain they give out as a prize would have to be shared with me. Although you can keep the trophy – I'm not an unreasonable man. I can see amazing marketing opportunities here. If horse racing is the sport of kings, then hen racing has to be the sport of Kentucky Colonels.

WORLD TOE-WRESTLING CHAMPIONSHIPS

I'll be honest with you, this one's just weird. It also sounds a bit unsanitary. You might catch athlete's foot off someone. Especially if they're a regular toe wrestler, which I suppose would make them an athlete. Or maybe you'd get a verruca. There are definite health and safety implications, is all I'm saying. There's an expression young farmers use when they go courting: 'Wrap it before you tap it'. I'm not going to spell out what that means in a family book, but I think it's definitely something to consider.

We apologize for featuring such frank and explicit images of unsafe practices in this volume.

Maybe some kind of dipping pool is the answer, like we use for sheep or cattle; or the little footbath thing you used to have to walk through to get to a public swimming pool in the old days. Although that was usually far worse than the pool itself – it had all sorts in it, things that no disinfectant could possibly kill. I wasn't surprised to learn that toe wrestling is one more idea that originated in a pub – Ye Olde Royal Oak Inn, in Staffordshire. Since then, it's moved to the Bentley Brook Inn in Derbyshire, in a village called Fenny Bentley, which I have to suspect was named by the scriptwriters for the *Carry On* films.

'It's called WHAT?'

I don't know if the people in Staffordshire were aggrieved to lose the toe-wrestling championship or glad to get rid of it. But I do know it can get pretty tasty, if anything involving feet can ever be called tasty. Three rounds per match-up: right foot, left foot, right foot again. And you can end up with serious bruising, or even broken toes. There's one thing I do like about it, though, and that is its GOAT – Greatest Of All Toes. He's a guy called Alan 'Nasty' Nash, who recently retired from the sport, aged 63, having won it 17 times. Not only did he work for JCB in his day job, meaning he was involved with diggers and so on (I love a bit of heavy machinery, as anyone who knows me can tell you), but he was also possessed of the mentality that typifies all elite competitors, and I consider him a role model. He once said, 'I'm useless at every other sport, so I'm really glad I found this.' Now those are the words of a true champion.

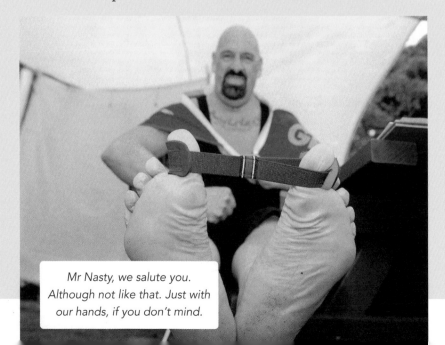

Mr Nasty, we salute you. Although not like that. Just with our hands, if you don't mind.

MAPLETON BRIDGE JUMP

You do have to wonder, don't you? I mean, farming is dangerous enough. But is there something in us country people that makes us say, 'Actually, you know what, farming *isn't* dangerous enough, I want to jump off a bridge thirty feet high into a freezing river, followed by a swim back to land and a foot race.' And it's Derbyshire, again, so perhaps the rest of the Midlands are looking at them and shaking their heads. I'm starting to suspect that Derbyshire is the county that, when it calls up the other counties and suggests going for a pint, is greeted by a set of hasty excuses because the other counties just know from experience how it's all going to end.

Call off the search. We've found Wally. He's living up to his name.

I have another theory, which is that, because farming is so carefully regulated – and so it should be, because it really *is* bloody dangerous, what with having to wrestle cows and pigs that could sit on you and squash you, or working with machinery that can chop you into little bits, grind you up or crush you – country people like the idea of something with no protocols a. Or at least none that will stop you killing yourself in some genuinely insane way. If there are any rules for something like this, they're probably all about the correct way to die. Keep your back straight, bend your knees, stuff like that.

'At last! The freedom to do something suicidally self-destructive without being obstructed by red tape. It's like Brexit, only quicker!'

I have another theory, too. All these mad, dangerous events invented by country people don't actually feature country people at all. It's just our way of luring city people to the countryside and making them do ridiculously risky things under the delusion that they're taking part in a countryside tradition. Whereas the real countryside tradition is watching them and trying not to show how hard you're laughing. Because no way would I do this – if my body was designed to go in water, I'd have gills – but I would totally watch it. 'Oh, yes,

'Welp, there goes another one. One of these days they're bound to twig, surely?'

you should definitely have a go. I'll just stand over here and see how you get on [*muffled cackle*]. Oh, you broke something? Never mind, it takes ages to get it right. Why not come back next year when you're healed up and try again?' Although you won't be able to. I'm sorry to say it, but this event has been discontinued. Not because everybody involved came to their senses, but because of issues over public access to the bridge – depriving all us potential spectators of some good, sadistic fun.

HARPOLE CLAY-PIPE-SMOKING CONTEST

Obviously, they do things differently in Northamptonshire. In Derbyshire, they're always inventing competitive ways to kill yourself quickly. In Northamptonshire, they've invented a competitive way to kill yourself slowly. No, this is not the Hyde Park 4/20, when all those stoners get together and puff loads of weed. At least with the clay-pipe-smoking contest you'd be able to walk past it without getting knocked unconscious by the fumes, then eating your own body weight in snacks when you come round.

I don't smoke, or do drugs. I don't get the fascination – the only drugs I have any time for are the ones you give to livestock to stop them getting ill. I hate tobacco smoke, so this definitely

isn't one for me. I like the leisurely aspect, mind you. The winner is the one who keeps their pipe going the longest. It started because people wanted to use up their tobacco before Lent. Like a smoking equivalent of Pancake Day. I'd take the pancakes every time, thanks, Lent or not.

> Yeah but smoking is cool tho... oh. Wait.

FOWNHOPE HEART OF OAK WALK

Now, this is a relief. Something completely sweet and wholesome. In this village in Herefordshire, instead of sticking decorated horse skulls on a stick, like that whole business of the Mari Lwyd in Wales (see page 62), where they want to give you nightmares for the rest of your days – which at that rate won't be many – they celebrate summer by decorating the sticks with flowers and leaves, then carrying them around in a parade. There's even prizes for the kiddies (and the adults) for Best Dressed Stick. It's basically mobile flower arranging. I think it's great that it doesn't have to cost you anything: everything you need, you can find in nature. It would never catch on round my way though. Good luck finding so much as a twig anywhere, let alone a stick. The village ramblers will have taken them to put in their log burners the instant they fall off the tree.

Aww. Thank you, Fownhope, you have made us want to live again.

KEELE CHRISTMAS-TREE-THROWING CHAMPIONSHIPS

Yes! Now we're talking. The only surprising thing about this is that it didn't happen sooner. Have you ever taken a Christmas tree out of your house? It's worse than Leonardo DiCaprio fighting the bear in that movie (*The Revenant*, if you want to check it out). You wrestle with it all the way. It falls on you, it cuts you, I swear it actually bites you somehow. No wonder by the time you finally get it out of there you want to chuck it as far away as possible. This is another one that's been recently discontinued, worse luck, but I hope they revive it. That said, there's no way I'm dragging my Christmas tree to Staffordshire. I might just have to hold my own contest here in Chipping Norton. Revenge will be mine.

'Take THAT, you vicious, needle-laden b*stard!'

MUCH MARCLE THE BIG APPLE

This is another one that sounds like the most countryside thing ever from the name alone. It's an apple festival in the village hall in Much Marcle, in Herefordshire. It's a great idea to begin with, but what I really like is that, at a nearby historic house, they have a sort of apple version of *Antiques Roadshow*, where you can take apples from your own orchard and the experts will identify the variety for you. When you think that we have 2,500 varieties in Britian, with names like Lady Pippinbottom and Spanking Rosy, this is bound to be fun. I want to go just so I can stick a bag of fruit on the table in front of the experts and say, 'Howdya like *them* apples?'

You had one job, apple festival. One job.

GEDDINGTON BOXING DAY SQUIRT

No way was I going to include this one when I heard the name. Until I found out that it's a gigantic water fight between two local volunteer fire brigades, standing on opposite sides of a river and trying to shoot a suspended beer barrel at each other using their firehoses. It looks an absolute blast, in every sense. I fancy rocking up with my slurry spreader and giving them a challenge. Ten bars of pressure, four thousand litres in the tank – that barrel will be over the hills and far away by the time they even get their faucets open. Give me five seconds and you'll never see it again.

Admit it. If you had the gear to do this, you'd never do anything else.

DERBY TUP

I wasn't going to include this one, either. What do they think it sounds like? Yes, it's a play, with songs, about slaughtering a huge sheep, but, you know, that's not what most country folk think when they hear the word 'tup'. Not. At. All.

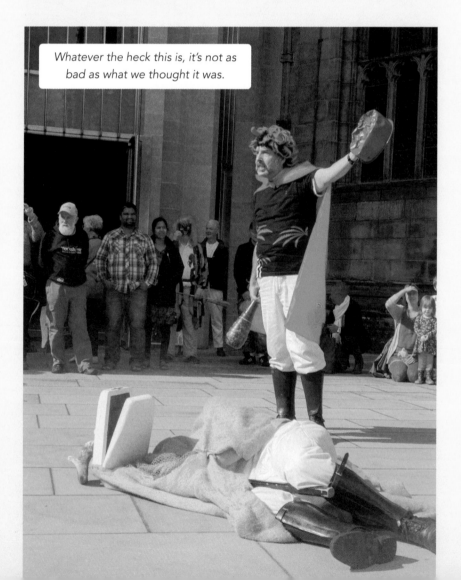

Whatever the heck this is, it's not as bad as what we thought it was.

POTFEST BY THE LAKE

They're just doing it on purpose now. I suppose you think you're being clever, Warwickshire? It's a ceramics festival, apparently. But bet you still get a few odd sorts turning up, assuming they can make it that far from Hyde Park in that state.

Michelangelo 0, Potfest 1

Chapter Six

North-West England

North-West England is one of those areas that people think must be really urban. It's got both Merseyside and Greater Manchester in it – which makes you wonder why it ever bothered going to war with Yorkshire. You'd have thought that might have kept it busy. I mean, they don't seem to like each other very much. If someone from Liverpool finds out I'm a Manchester United fan, they always say something like 'of course you are, you're from Oxfordshire'. But almost every time I see someone in a Liverpool shirt, they turn out to be from Wales, or Norway, or Tierra del Fuego. When aliens finally land, I'm betting the first thing they'll do is go to either Old Trafford or Anfield and take the stadium tour, and the second thing they'll do is go to either Anfield or Old Trafford and take a selfie sticking up two fingers at it.

'COME ON YOU GREEE . . .
Reds, I mean.'

I suppose the Wars of the Roses took place before all of that, when they didn't have things like the Industrial Revolution and football rivalries to distract them. They must have been bang into their gardening though to fight about something like that. Mind you, I've been to a few flower shows and I know how competitive it gets, so I can see how it would go. Mrs Arkwright makes a catty remark about Mrs Walmsley's floribunda and the next thing you know you've got two armies facing off across Hedgeley Moor because *nobody* talks like that to the missus.

'Go on, say that thing about aphid damage to my Blush Excelsa again – I bloody dare you!'

Fair play to them, they don't let it go in a hurry. Most other regions' traditions and contests seem to centre around booze, one way or another. Which is no surprise, because you need to be hammered to even think about most of them, let alone actually do them. But in the North-West, it's mainly about two things: food, and taking the p*ss out of Yorkshire. There's a food theme to everything. Just like down our way, we've got The Wurzels, who call themselves a 'Scrumpy and Western' band, so up there, they've got The Lancashire Hotpots, who have songs like 'Chippy Tea'. I like the single-mindedness of that. In the North-West, they've decided what matters, and they stick to it. Or rather, it sticks to them, unless they use a napkin.

Anyway, there's plenty of the North-West that's very rural –

How can you eat so much of that …

lots of moorland
and heath, lots
of farms – and
even the parts
that have been
built up are still
a bit partial to
the old-school
grub and the
old-school
traditions, so

... and keep those waistcoats so clean?

good for them. And they love a flat cap, too, which I totally
approve of. Farmers and Northerners are the flat-cap hipsters
– we were wearing them long before gangsters in Birmingham
ever made them cool.

'I'll Peaky Blinders you in a minute, my lad.'

RAMSBOTTOM WORLD BLACK-PUDDING-THROWING CHAMPIONSHIPS

This is a perfect example of what I was talking about: a custom that combines heavy-duty foodstuffs, and poking Yorkshire with a stick. It's great that it happens in a place with the most Lancashire name I've ever heard. Once again, I love the way that so many of these mad contests stick the word 'world' in their name. As if they want to make sure you don't mix it up with the black-pudding-throwing championship in, I dunno, Osaka. Then again, the only people who like mad contests more than us British are the Japanese, and they love doing British stuff too, so maybe there's a good reason to have 'world' in there somewhere.

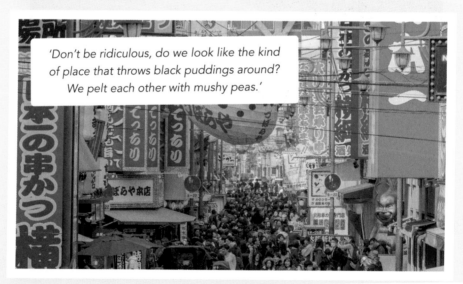

'Don't be ridiculous, do we look like the kind of place that throws black puddings around? We pelt each other with mushy peas.'

'You rang?'

I've got to say, I like black pudding. It's one of those things, like brawn, that is delicious but you don't want to think too hard about what went in there. It makes you wonder, though, who was the person who first came up with that idea? Who looked at those ingredients and thought, yeah, let's put them together and see how it comes out? Because you wouldn't go near any of them on their own, would you, unless you were some kind of pig vampire.

What happens in Ramsbottom is they stick a load of Yorkshire puddings on a platform twenty feet above the ground, then you have to try to knock them off with a black pudding. The story goes that at one point in the Wars of the Roses, the two sides ran out of ammunition and started chucking their dinner at each other. I'm no historian, but that's a chinny reckon if ever I heard one. It's obviously another idea that somebody thought of in the pub. Sort of like *Angry Birds*, only with puddings instead of birds and pigs. What's clever about it, though, is that if you're from Yorkshire, you can't turn it around. Try knocking down some black puddings with a batch of Yorkshire puds and see how far you get. I'm all for the whole thing – you work out a bit of regional aggro without anyone getting hurt. No battles, no arguments about floral arrangements, nobody gets their head put on a spike, none of that. The only thing is that, like a lot of these competitions, it does seem an awful waste of food. I do love a Yorkshire pud, but it needs a lot of gravy, and I don't see any involved here. Maybe if you had to knock them down into a huge vat of gravy that would improve the whole business. But then a huge vat of gravy tends to improve anything.

"'Ave some of that, y' White Rose muppets!'

STACKSTEADS WORLD GRAVY-WRESTLING CHAMPIONSHIPS

I take it back. A huge vat of gravy does not, after all, improve everything. I don't watch a lot of WWE or any of that stuff, but if I do, I don't find myself thinking, well, this is OK, but what they really need to do is slather the wrestlers in Bisto. Although don't tell the people behind this event I said that, because I'm sure they'd never descend to using gravy granules from a shop. They'll descend to a lot of other things – not excluding writhing around in a sixteen-foot pool of hot, brown liquid, grappling with each other – but never that. They're proud of their gravy in Lancashire, and no doubt with good reason. Bit of a funny thing to do with it, though. If I was writing a list of 'things to do with gravy' it would have one item on it and this would definitely not be it.

It's a logical impossibility yet somehow this manages to be simultaneously savoury and unsavoury.

Don't get me wrong, I'm not averse to a food fight. It can be good fun. I spilled the beans, so to speak, on potato fighting in my last book, *The World According to Kaleb*, but this has to be a step too far. The worst you get from a potato is a bruise, maybe for you, maybe for the potato, but gravy's going to get everywhere. Every single wrinkle and crack and … things that should be free of gravy generally. You'll be whiffing of the stuff for weeks afterwards. Dogs following you in the street. The cows looking at you suspiciously whenever you go to feed them. You'll be a human stock cube, basically. Your bathwater will be like soup. I like that this event is for charity, the East Lancashire Hospice. But, you know, anyone else would just have a raffle. Still, credit to them for thinking different. Also, you can get hosed down afterwards by the local fire brigade. I bet it doesn't get the worst of it out, but at least it's a start.

Right, action stations!

WIGAN WORLD PIE-EATING CHAMPIONSHIPS

Just in case you thought I was exaggerating with the whole food thing. I mean, the whole thing about food, not a thing about wholefood. The only way a pie could be called a 'wholefood' is when you eat the whole pie. Which I'm all in favour of, as you only have to look at me to know. I'm not into competitive eating, though, unlike some of my mates. They train for it. I went for a steak with a few friends the other day, and this plate arrived with five steaks on it, mushrooms, two bowls of chips, you could feed a family with it, and one of my mates challenged me to eat the whole thing in seven minutes. Not a chance. And he did his in four. I don't even know why. It wasn't a competition. I like to take my time over my food, have a chat, eat a bit, have another chat. So I don't think I'd be any good at this. I'd still be tucking into to my first pie and asking the person next to me how their barley was coming along, when they were onto pie number seven.

'The pies are all right, but what I really come for is the sparkling dinner table conversation.'

Pie just out of shot (right).

How it used to work was the winner was the one who ate the most pies within a certain time, thus answering a time-honoured question you hear sung at football matches. But now they've changed it to meet government healthy eating guidelines, and you get just one pie, twelve centimetres across and three and a half deep, and the winner's the one who eats it fastest. The current record is 23.53 seconds (well done, Mr Martin Appleton-Clare). That's just the government all over, isn't it? We can't have anyone having any sort of fun, can we? Oh no, we'll put a stop to that. Healthy eating guidelines and pies just don't mix. That is the whole point of pies. And now a magnificent spectacle, a kind of epic Pie Marathon, has been reduced to a Pie One Hundred Metre Sprint. My favourite bit about the whole thing is what you get if you win: a free meal. Because you're really going to fancy that after, aren't you?

Float like a butterfly, roll like a pie.

WYBUNBURY FIG PIE WAKES

In this Cheshire village, they bake fig pies with a special hard pastry, then roll them down the street. I bet I know how this one got started. Two hundred years ago, somebody made a batch of inedible fig pies, then pretended they'd done it on purpose, and everybody joined in chucking them down the road so they wouldn't have to eat them. It's like when you're a kid and you have to pretend you like your dinner while craftily trying to give it to the dog. Today it's a serious contest, with different age classes and everything. I can see the montage now. The hero training. Concentrating. Meditating. Getting a massage. Then the big finale, when he wins. If I did this, I'd come in to a theme tune, like a boxer. 'Colt 45' by Cooper Alan & Rvshvd. Stare out my opponent. Start rhyming about how I'm too pretty not to be the champion, and my perm's too good not to win. I AM THE GREATEST.

BURY PACE EGGERS

The surprising thing about this one is, despite what it sounds like, it's not about food, for once. They're a group of mummers – actors who perform traditional plays during Easter, which is where the egg part comes in. There are a lot of these around the country, but what makes the Bury one stand out for me are their character names: Tosspot, The Doctor, Big Head, Beelzebub and The Fool, Slasher The Turkish Knight. It sounds like all the people who've been banned from your local pub and had their names written

up on a piece of paper stuck above the bar. 'Slasher The Turkish Knight? No, he's definitely not allowed in, not after what he did last time.' Maybe the Eggers got together in the first place to do a play about how they all got chucked out *FOR NOTHING*. Because it's always for nothing, isn't it? 'Wasn't me.' But I saw you. 'Not me.' *Everyone* saw you. 'Nah, I didn't do anything.' We've got signed witness statements. 'Wasn't me.' And CCTV footage. 'No, not of me.' And photographs. 'Wasn't me.' It's that Shaggy song come to life. So I'd definitely watch the Bury Pace Eggers, if only to see how much they remind me of a typical night down at The Red Lion.

'Leave it, lads – let's go and start a traditional Lancastrian street theatre troupe instead.'

MOULTON CROW FAIR

I had to mention this, because it's all been quite jolly and food-related in the North-West so far, then along comes something almost as terrifying as those Welsh horse ghosts. Crows freak me out. They're well scary. And they're clever, too. Evil, and clever, which is a bad combination. I've seen them peck the eyes out of sheep, and break into things to get to food, and you do not want to mess with them. So the idea of going to Cheshire to see a bunch of giant crows dancing around is, frankly, bloody petrifying. It's like *The Masked Singer* – strictly anonymous, you have to stay in costume – but how they don't frighten each other to death, let alone the bystanders, I have no idea.

'AAAAAAAAAAAAAH!'

Chapter Seven

The Scottish Highlands

The main thing I know about Scotland is that the country is unbelievably beautiful, and the people are unbelievably tough. I've definitely got that the right way around. I've met a few Scottish farmers, and I'll tell you, they could plough a field full of rocks by sun-up, then eat the rocks by sundown. Every time I see anything about Scotland, they always have the most amazing pictures. The scenery is just spectacular. It's funny, the number of Scottish farmers I meet, at the farm shop, for instance, who all say they love it down here. And I tell them I love it up there. I suppose it's literally a case of the grass being greener on the other side of the fence, if you can call Hadrian's Wall a fence.

Gotta say, it looks pretty much the same shade on both sides to us.

'We city people have our own amusements, you know, even if they're not quite as sophisticated as hurling great bits of lumber across the landscape.'

There's two main parts of Scotland: the Lowlands, where there are around five million people, and most of those are in a few cities, so there's loads of country land around them; then there's the Highlands, which makes even the empty bits of the Lowlands look like rush hour in Mexico City. Only with not as many sombreros, I suppose, unless they've just come back from holiday. So, it's properly rural. In fact, nowhere is more rural than the Scottish Highlands. They've got pretty much the lowest population density in Europe. Rural people basically think of fun completely differently to city people. Lots of their festivals and traditions are properly rural – real farming games, things that no city person would ever think of, let alone do.

UP HELLY AA

They've got a lot of Viking festivals up north, but none of them is quite as spectacular as this one. It's in Shetland, which is almost as close to the Faroe Islands (somewhere I know exists because their football team turns up on the telly occasionally), and to Norway, as it is to the Scottish mainland. So, when other places celebrate their Viking heritage, I tend to think, are you sure? Or are you just doing a glorified stag do, so you can romp around wearing helmets with horns on them and drink really strong lager out of plastic skulls and all that? (At least, I hope the skulls are plastic.)

It seems like everybody claims they have Viking heritage. Except in Banbury, just up the road from me. I met a bloke who told me the Vikings never got to Banbury, and that's why Banbury has the lowest average IQ in Britain. I have no idea if either of those things are true. Although I've been to Banbury, and I'm not 100 per cent sure he's wrong about the second one. (Joking! Love you, Banbury!) In fact, come to think of it, this bloke might have been from Brackley. No love lost between those two towns.

But when people in Shetland say they're descended from Vikings, I'm not about to argue with them, because they basically live on a rock off the coast of Scandinavia so: (1) it is probably true; and (2) if it isn't, I'm not going to be the one to tell them. Anyone who lives on a rock off the coast of Scandinavia is bound to be absolute nails and they can say they're descended from Zulus if they like and I'll just nod along. Also, their festival looks like a lot of fun. On the last Tuesday in January, they all dress up like a Dark Ages version of ABBA, and a guy called a Guizer Jarl leads them around the town of Lerwick, pulling a galley, which is another name for a Viking longboat. At the end of the whole thing, they set fire to the galley. If we did that round here, we'd catch hell for polluting the atmosphere, but I suppose smog isn't that much of an issue when you're bang in between the North and the Norwegian Seas. I like that they've got place names there like Ramna Stacks, which sounds like an old-school wrestler, and

Yell Sound, which is probably what you did when you saw the Vikings arriving. It's a shame about burning that galley – if you put a motor on it, it would be worth a lot of money. You could take it to London and make a fortune carrying punters up and down the Thames for a Viking party. But tradition is tradition, and burning stuff is the Viking brand, so fair play to them for sticking to it. Also, it's in January, and it's up near the Arctic Circle, so it's probably the only way to keep warm. Let's be honest, we all like seeing something burn. Especially if it's something really big, it belongs to someone else, and you're the one who got to set light to it.

We're not saying this is the most exciting thing we've ever seen, but we have to admit, this is the most exciting thing we've ever seen.

ST MARGARET HOPE'S BOYS' PLOUGHING MATCH AND FESTIVAL OF THE HORSE

The more I learn about Scotland, the more impressed I am. When I heard they have a ploughing contest for kids, I thought, brilliant. Ploughing is a dying art. In a few years' time, nobody will know how to do it. Except in Orkney, which is so far north that even Scotland thinks of it as northern. I was so excited about this I wanted to enter my lad Oscar into it, and I even started putting the route into the satnav, until I found out that Orkney is a bunch of islands, so – until I get hold of James Bond's amphibious car – that one's out the window. (I may not know who James Bond is, but I know a good bit of kit when I see it.)

'Welcome to Orkney, Mr Bond. We've been expecting you.'

'You missed a bit, sonny!'

It turns out it isn't just a ploughing contest for kids – it's a contest where the kids actually power the ploughs, miniature ones. They have forty-five minutes to plough a four-foot-square patch of ground.

So now I'm not so sure. On the one hand, I think it's amazing they're passing on these skills. That's what we all need to be doing on a Sunday. And I'd be happy if my boy grows up to be a farmer. Just not so sure I want him to grow up to be a horse. And if you think I'm worrying for no reason, they dress lots of other kids in costumes that represent horses. It's not as bad as those insane Welsh creatures of nightmare. In fact, it's really quite sweet. Come to think of it, Oscar would look really cute in that. I think his mum would love it. What the hell, I'm back in. Let the miniature ploughing commence!

Neigh, lads

WORLD STONE-SKIMMING CHAMPIONSHIPS

Aw, this is great. I love skimming stones anyway – it's the most relaxing thing you can do – but imagine doing it here, on Easdale Island in the Hebrides, in a ridiculously beautiful setting, on the calmest water. Just look at the place. Seriously. Just look at it. You could go there to skim stones and it would be so peaceful you'd never bother going back home. Why would you?

Even the lack of tractors can't put you off.

I wouldn't even care about winning, and I always care about winning. I'm so competitive that I'd race you to see who could tie their shoelaces faster. It's not the taking part that counts, it's showing your opponent who's best, which is me. But I'll make an exception for this. I don't mind if somebody beats me. They won't, though, because I'm really good at skimming stones and I'll practise non-stop for at least a year beforehand.

Close, but no cigar.

HAGGIS-EATING, HAGGIS-HUNTING & HAGGIS-HURLING

I've done my best to find out about haggis. (Or should that be haggises? Not sure if it's like sheep, where more than one is still sheep, unfortunately – or like cows, where more than one is, well, cows.) The trouble is, I've got conflicting information. One source says that a haggis is a traditional Scottish dish made from a sheep stomach filled with minced and spiced

'Either way, we've caught one!'

offal and oats. Another says that it's a small, elusive animal, possibly furry, possibly tartan-coloured, with legs shorter on one side than the other to make it easier to run around hills. But who to believe?

I suppose that makes haggis-hunting a Highland sport. And the World Haggis-Eating Championship takes place at the Birnam Highland Games, and is free to enter, so I'm tempted to go to that, even if it means eating bits of sheep I definitely don't want to think about.

The haggis-related event I like the most, though, is haggis-hurling. The World Haggis-Hurling Championship takes place at the Bearsden & Milngavie Highland Games, which are held near Glasgow, and are actually in the Lowlands. All of which makes me think that Scotland is perpetrating an epic wind-up. Maybe they got the idea from Wales. But two can play at that game. The sport was revived in 1977, based on an ancient custom where Scottish women would throw haggis – haggises – haggisisisises – whatever – to their husbands while the husbands were doing whatever ancient Scottish husbands did for a job, and the husbands would catch the haggohsuityourselves in their kilts. Now they have contests all over Scotland, as well as in Canada and Belgium. Then in 2004, an Irish guy admitted he'd made the whole thing up as a hoax to test how gullible Scottish people were. No way am I going to get in the middle of an argument between Irish people and Scottish people, thanks – I want to live to see my kids grow up,

Possibly the only competitor in the haggis-hurling event who is younger than the sport of haggis-hurling.

or just next week, come to that. So, point me at my haggis and show me where to throw it and I'll get on with that and leave you to sort out the details.

It's given me an idea, though. If anyone can come in and persuade Scottish people that some daft sport they've invented in the pub is actually an ancient tradition, why not me? So, I've come up with the Highland Cow Ring Toss. It's a bit like a fairground game, where you throw hoops at a target, but here you have to land them on a Highland cow's horns. If the cow moves, well that's just part of the game. So, being as competitive as I am, I'd bring my own cow that I'd trained to stand really, really still.

We think this might be cheating tho.

STONEHAVEN FIREBALLS

This is crazy. Stonehaven's a town on Scotland's north-east coast, and the way they like to see in the new year is to make balls out of wood and cloth, attach them to wires, soak them in paraffin, set fire to them, and get forty-five of the people who live there to march through the streets whirling them around their heads. I mean, you'd want to get them to sign a legal disclaimer before doing that. Health and safety, nil, making your own entertainment, ten. It sounds dangerous, it is dangerous, and I want to go and see it as soon as I can persuade anybody to sell me enough insurance.

We understand Jerry Lee Lewis had
some thoughts on the matter.

Chapter Eight

The Scottish Lowlands

While we're still on Scotland, I think my name originates there. Somebody once told me it's in the Scottish Bible. I didn't know there was a Scottish Bible. I'm not even sure what that means. Maybe it's got a tartan cover or something. Or perhaps it's a reference to the famous King James Bible, an authoritative and epochal translation into English by a panel of 47 Biblical experts, commissioned in 1604 by James I of England who was also James VI of Scotland, in which the Old Testament refers to Caleb, the companion of Moses, who helped bring the Children of Israel into the land of Canaan ... sorry, I'm not sure what happened there, I went all dizzy for a minute and when I came round those words were on the screen.

Caleb offers his daughter, Achsah, in marriage to the man who conquers Kiriath-sepher. As you do.

CLIMBING THE GREASY POLE

'Not to worry, I was eight feet tall to start with, I can afford to lose a bit.'

Anywhere else in the world this would be a figure of speech for my approach to life. You know – a metaphor for getting on, working hard, aiming for the top, achieving success in difficult circumstances.

Not in Scotland, though.

In Irvine, Scotland, they cover a thirty-foot pole in axle grease, then climb up it. They're not just a hardy lot, your Scots, they also seem to be quite literal-minded.

It's not only one person at a time, either. It's a team sport. You need four people. And they basically need to be stackable. So, you need one really solid, heavy-duty guy at the bottom, two lighter ones to stand on him, and a small, nimble feller right at the top to wriggle up the pole. If I had to do it, I'd definitely rather be the one at the top of the pile – and it usually is a pile, by the time they're finished.

Some years nobody makes it to the top of the pole, but the guy at the bottom comes out a few inches shorter.

I have to admit, I'm not really qualified for any of those roles. I'm not built for climbing, any more than I am to be a jockey. I'd crush anyone I tried to stand on. But I'm not exactly equipped to be stood on, either. I was made to be on the ground. Solo. Or on a tractor, at most. And I am highly accident prone, which is bad enough when you're a farmer, and a hundred things are waiting to kill you before breakfast – a hundred and one if you include breakfast itself – but is probably even worse when you're a greasy-pole-climber. I can fall off my floor, given half a chance, so I have no desire to tempt either fate, or gravity. I would much rather remain on good terms with gravity, so it doesn't decide to hurt me. I am designed to be ground-adjacent, and that's how I want it to stay. I definitely fancy going to Irvine to see this, though. It looks like a brilliant spectator sport. Plus they put a ham at the top of the pole as a prize. Maybe I could sponsor a team – Team Kaleb – in exchange for a share of the ham. It would probably be cheaper just to buy the ham, but it would be a lot less fun.

It tastes so much better when you've really earned it.

SCOTTISH ALTERNATIVE GAMES

Everyone knows about Scotland's Highland Games, which seem to consist mainly of chucking really heavy stuff around. As a farmer I can totally relate to this, although I draw the line at seeing how far I can throw a bloody great tree, which they call 'tossing the caber'. I can't lie, if you said that phrase to me in the pub, that's not what I would have expected it to mean, but this is a family book and I'm not going to, um, expand on that any further.

You never hear anyone mention the Lowland Games, though, do you? In New Galloway, they decided to do something about that, and came up with the Scottish Alternative Games. Which

Wait, what?

are definitely a bit less hardcore than all the Highland Games stuff like caber toss, stone put (like the shot put, only with a stone that weighs almost *two* stone), hammer throw, and something called maide leisg, where you have to pull your opponent up off the ground with a stick you're both holding. No, the big event at these alternative games is gird 'n' cleek – which is a bunch of kids propelling a hoop along the ground with a stick.

I won't deny that's a difficult skill – a bit like rolling a tractor wheel along the ground without it falling over, and I can tell you, that's no doddle. But still. It doesn't make me think of tough Scottish people so much as Victorian urchins.

'Oot ma road, lassie, ah'm off tae win the gird 'n' cleek at the New Galloway Alternative Scottish Games.'

It's the other events they've got that really interest me. Tractor-pulling, for a start. Now that's my kind of thing. Even before I knew what tractor-pulling was, I knew it was my kind of thing. Was it pulling a tractor, or pulling something with a tractor? Or maybe a tug of war between two tractors? That would be fantastic. Whoever's got the best traction grip and the most horsepower wins. (Hint: it's going to be me.) They've also got snail racing, which could be a bit tedious, or could be brilliant, depending on how close it comes to how I like to picture it.

I don't think snail racing will do it for me, though, because I'm very impatient. But so would you be, if you had to work with Jeremy. He's trying to plough a field, and I'm all, why don't you do it this way, why don't you finish it quicker? I'm like the Roy

'Oot ma road, all of youse, ah'm off tae win the snail racing at the New Galloway Alternative Scottish Games.'

Keane of ploughing. But then every farmer's like that. It's their way, or no way.

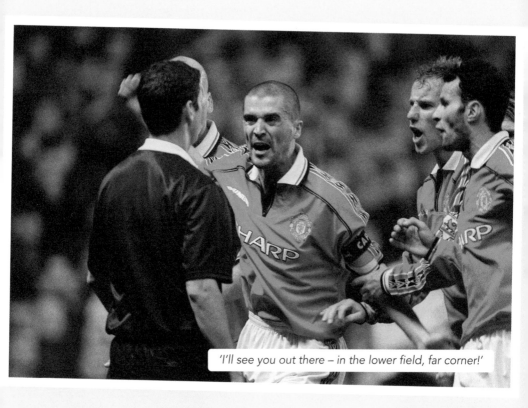

'I'll see you out there – in the lower field, far corner!'

Then there's tossing the sheaf over the high bar. I got very excited when I first heard about this, because I misheard and thought it was sheep – and you can't toss sheep far enough or high enough for my liking. I'm joking, obviously, I hate animal cruelty, but that's why I hate sheep – because if you know anything about sheep, you'll know that nobody is more cruel to sheep than sheep are, plus it's animal cruelty towards me, just having to be anywhere near them. Anyway, this is a sheaf – a burlap bag filled with straw, weighing sixteen pounds, and you have to toss it over a horizontal bar above your head with a pitchfork. I reckon I could get a team of Oxfordshire farmers together and have a decent go at that one. It still sounds like a good idea to me to do it with sheep, though. OK, maybe without the pitchfork. You could do it before shearing, so they've got natural padding and protection, and prepare a soft landing area, so they just come down with a *flump*.

'Nice landing, son!'

CARNWATH RED HOSE RACE

OK, this is just a three-mile race, and there's nothing amazing about that. Although I have to be honest and admit it would be amazing if I could run it without dying, but as the old song says, I was built for comfort, I wasn't built for speed. There are two things about it that make it worth a mention. One is that it claims to be the oldest surviving foot race in the world, dating back to 1508. So, what I want to know is, how? How on earth do they know that? That's just one of those things you can say about anything, because who's going to prove you wrong? Still, with my marketing head on here, I'm impressed. That's one way to draw attention to what's really just a historical fun-run. The other thing is the prize, which is how the race gets its name. If you win, they give you a pair of socks. Red ones. But still socks. I get nothing but socks for Christmas. Every year. I'm not gonna go to Scotland and knacker myself on a three-mile run just to get some more.

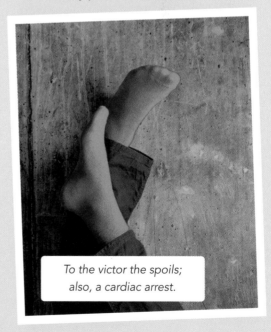

To the victor the spoils; also, a cardiac arrest.

MOFFAT SHEEP RACE

I think I've made my feeling on sheep races pretty clear already. I didn't think it could get any worse. But how wrong I was. If you take that mad Spanish festival where they let a load of bulls run through the city, and you do it with sheep instead, you'd get something like this. The difference is, instead of the bulls trying to kill you, you'll get the sheep trying to kill themselves. It must be like some kind of mass sacrifice. They'll be in the road thinking, 'OK, where's a car that I can throw myself under? What if I run headfirst into a lamp-post?' It's not an ancient tradition or anything – they've only been doing it since 2011. Still, you have to give Scotland credit for being forward-looking. It's basically a way to give new opportunities to suicidal sheep.

'Quick, girls, I think I see an open manhole
up ahead that we can fall into!'

*Well, that's us trying to stay awake for the next fortnight
in case this somehow gets into our dreams.*

QUEENSFERRY BURRY MAN

I don't want to be rude or anything, but what the absolute f*** is this?

I didn't know these things were called burdock heads. But I know all about them. Every farmer does. Burrs. They get everywhere. On your clothes. On your livestock. On your dog. And just try getting rid of them. You can spend literally hours picking them off and swearing. So, anyone else would never want to see them again. But not in Queensferry. Here, they keep them and make a huge, heavy suit out of them. Then they dress some poor sod in it and make him walk around town. He looks like the world's creepiest superhero, and honestly, I think if he turned up to save me from impending doom I'd rather just take my chances with the impending doom. He has to have sticks to hold his arms up to stop them getting stuck to his sides, and two helpers to get him about. As if that wasn't mad enough, he has to walk for seven miles around the town while the people give him whisky whenever they see him, for good luck. I don't know whose. Definitely not his. Maybe they're hoping that they get lucky and don't have to wear the suit next year. He has to sip the whisky through a straw because he's also wearing a balaclava covered in burrs. No wonder the poor guy needs a drink – or fifty.

THE LOONY DOOK

This isn't that black and yellow cartoon character who gets angry and spits a lot, and who definitely isn't Scottish, even though he could probably pass for it on a Friday night if he had to. He looks bang up for a row, is all I'm saying.

No, what this is is, on New Year's Day, loads of people in Edinburgh put on fancy dress and go and jump in the Firth of Forth, which is one of those things Scottish people have just to confuse everyone. You know: 'The what of what?' 'Firth of Forth, pal.' Well, now you just sound like that cartoon duck in a maths class. Down here, we'd call it an estuary, not that I've ever been near enough the sea to see one. Up there, I'd call it absolutely bloody freezing. But the Scottish don't believe in hypothermia. That's for

'Put 'em up, you thhcoundrel! I find your thhtereotypical humour dethhpicable!'

148

soft English types. They think it makes a good hangover cure, and I suppose it does – you're not going to be thinking about your headache when your feet fall off, are you? Definitely kill or cure, that one, and most likely kill, I'd have thought, but they seem to thrive on it. That's how hard the Scottish are. And bear in mind this one isn't even in the countryside. This is what Scottish *city* people are like. So just imagine how rugged they must be in the country. Nothing but respect from me there.

'No, we did not lose a bet, and nobody is holding our families hostage. Why do you ask?'

Chapter Nine

East England

This is the flat bit, isn't it? When I meet farmers from other parts of the country, no matter where, I always hear the same thing: 'You're farming boys' land – we farm man's land!' Well, maybe, in some cases, but in East England, the fields are all nice and even. I reckon, let's switch farms for a day and see how you like it. You try farming on a twenty-degree slope, then come and say that to me. Their idea of a mountain is called the Isle of Ely, because it's about two feet higher than anywhere else – still completely flat, but the only bit that isn't underwater when it rains. In fact, twenty per cent of the whole region is below sea level, so it's a good thing most of those bits aren't closer to the sea.

'Let's see you plough this, then.'

I do like East England, though, because it's really traditional. It's a bit like Scotland in that you've got most of the people in a few urban areas and then there's lots

Not being funny or anything, Essex, but if you can look out your window and see this, are you really rural?

of space. If you take Essex out – which I reckon you should because (no disrespect to the proper country bits of Essex, and there are lots of them), most of it is basically London now – you'd only lose a fifth of the area but almost a third of the people, which tells you something about the rest of East England. Nice and empty, like the countryside should be, so we can get on with the business of farming it without people getting in the way.

BRIGG FAIR

When I first heard about the Brigg Fair in North Lincolnshire I wasn't too impressed initially – what kind of country event doesn't open in the morning? But then I heard that it started in 1205 – which makes it eight centuries old. Eight. Centuries. Old. Think about that for a minute. Or a century, if you like, the Brigg Fair won't care either way. I don't know what my ancestors were doing in 1205, but I bet it wasn't gaining a charter to run a horse fair. This is a big deal to the travelling community, who come to show off and trade their horses,

Love the feathering over the hooves, but it must wick like a pair of flares.

which they're very proud of. And I can see why – even though the grounds aren't all that, and sometimes they're muddy, the horses always look smart. You get some of the best horses in the country there.

I don't have a horse, but I totally get why the horse people love the horse fair, because I'm the same with tractors. Just thirty or so miles away over the county border with Nottinghamshire, there's the Cereals event, which is one of the biggest agricultural shows in the country. They hold it in an incredibly flat field, so you could be in East England for all you can tell. They have loads of tents and, because there's nothing to stop the wind, you think they're going to blow away at any moment. And they have a ring where they do massive tractor demos – both the demos and also sometimes the tractors are massive – as well as lots of other serious agricultural machinery on display, so obviously I love it. If you're a tractor enthusiast – obviously, I am *the* tractor enthusiast – this is the show for you. Tractors are to me what horses are to the travelling community. They're therapy machines. Every farmer has a tractor they just can't get rid of, even though they probably should. It's too much a part of their life. You might sell the others, but you'd never sell *that* one, and I bet you the horse traders all have a horse like that too. Horses are expensive to keep – tack, trim, feed, all that. But you never have to change a tyre on one, and they don't have electrical breakdowns, so swings and ladders, really.

*Not gonna lie, we don't even know what
this stuff does, but we are in awe of it.*

WHITTLESEA STRAW BEAR FESTIVAL

Honestly, I have no idea what is going on here or why. All I know is that they do it on Plough Monday, which is the traditional day when farm workers come back to work after Christmas and parade their agricultural equipment – I like the sound of that. The straw bear costume weighs five stone. Imagine walking around town all day in that. I mean, I know it's January, but you'd sweat to death if the weight didn't kill you first. Plus, you'd have kids trying to set you on fire, and me chasing you through the streets with my baler, just waiting for my chance. Don't envy them at all.

'Let me take you down 'cos I'm going to Straw Bear-y fields...'

SWATON WORLD EGG THROWING CHAMPIONSHIPS

This looks excellent – if anyone makes the world's worst pun about that opening, I'm going to use them as a target. I can see myself getting serious about this one, looking for the perfectly shaped egg: not too long, not too round at one end, not too pointy at the other. There's something about holding an egg in your hand, isn't there? You could be an eighty-year-old vicar but right away you turn to the dark side: your mind is filled with evil thoughts, and you're like a naughty schoolkid, thinking how you could cause the most mischief with it. Even if you're just cracking it, or putting it in a hot pan, you have to battle the temptation to cause havoc.

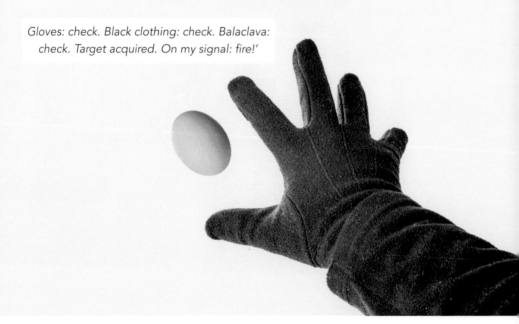

Gloves: check. Black clothing: check. Balaclava: check. Target acquired. On my signal: fire!'

Mau! Mau! Di-di mau!

I like that this is a serious competition, with some carefully thought-out games, rather than just flinging a load of eggs at the nearest person and seeing how many you can land – although, let's be honest, that would be amazing fun, too. (Hang on, I've just checked, and you *do* get to do that, plus you get extra points for hitting especially sensitive bits of the target's body.) For the main event, you put two people ten metres apart, and they have to throw a raw egg from one to another. If the egg survives, they keep moving further apart until it breaks. The ones who manage the biggest distance win. Plus they've got an egg trebuchet (had to look this word up – it means a machine used in medieval warfare. You learn something new every day!), which is obviously going to be one of the greatest inventions in the history of humankind. And my favourite part: Russian egg roulette. I thought Russian egg salad was dodgy enough, but this is on a whole other level. One half-dozen box of eggs: five hard-boiled, one raw. You pick one and smash it on your forehead. Not sure if spectators bet on you while you're at it, but if not, they should do.

OASBY BABOON-TOSSING NIGHT

As I keep having to repeat, this is a family book – and I'm trying to keep it that way – but some of these traditions don't make it easy.

Fortunately, this is another one that's not as bad as it sounds. Just how traditional it truly is depends on who you believe. The event was supposedly inspired by a tragic incident in the early eighteenth century, when a pet baboon threw the baby son of a local landowner out of the window of Culverthorpe Hall. Surprisingly, this didn't lead to a Dangerous Primates Act being debated in Parliament; however, almost two hundred years later, it did lead to a bunch of Morris Men in a Lincolnshire pub (another bunch of men in a pub, isn't it?) deciding it would be a good idea to commemorate it by dressing a guy up as a baboon and chasing him through the town, because, well, obviously. Then they all go back to the pub, but the landlord won't let them in until somebody succeeds in chucking a toy monkey over the pub roof. Then everybody, including the baboon, piles back and gets (even more) p*ssed. Fair enough, I can think of worse ways to spend an evening, and I quite fancy testing my stuffed-monkey-hurling skills, so, if I ever get round that way, I'm in.

And to think people take the mickey out of Norfolk when this is going on up the road.

OLD BOLINGBROKE
CANDLE & PIN AUCTION

This is a genuine old tradition, the kind of auction that used to happen a lot. Boom! Nowadays, Horncastle in Lincolnshire is one of the few places that still does it. Each year, in late March, they auction off the grazing rights to a six-acre field until October, so you get a real farming crowd in attendance. They then stick a pin through a candle, light the candle, and the auctioneer accepts bids for the grazing rights until the moment the flame melts the wax around the pin and the pin falls out. Basically, somebody back in the old days invented eBay before there was even the Internet. Must be a proper battle of nerves, this. I know what farmers are like at auctions. I went to one the other day, and there was a cow that everyone knew was worth a thousand pounds, but they're still upping their bids by a fiver a go. So this is a good way to stop them wasting everybody's time.

They've got a similar idea an hour away in Bourne, which is also in Lincolnshire. (Lincolnshire is a pretty big county, it must be said, and, as a rule, the bigger the county, the more rural it's going to be.) They call it The Whitebread Meadow Running Auction, and it lasts as long as it takes two kids to race a two-hundred-yard dash. I say 'dash' but – well, let's just say it's a good thing it's not for anything bigger than some grazing rights, and that the money goes to charity. Because if I was selling anything in this way, I'd make sure to pick the two slowest children in the area. Basically, anyone who reminded me of me as a kid.

You might not consider this the height of excitement but it's a lot more thrilling than hunching over your computer while drunk in the small hours trying to snipe a last-second bid on a VHS box-set of a sitcom you're not sure you ever really liked when you don't even own a video player.

WITCHAM WORLD PEA-SHOOTING CHAMPIONSHIPS

One thing I like especially about the events and customs in East England is that there's something really gentle about most of them. Everywhere else it seems people want to find mad ways to maim or kill themselves, or each other, or at least be bloody terrifying (I'm looking at you, Wales). Here, they've got not one but two events that are straight out of the *Beano*. We used to have spitballs at school, where you'd shoot little bits of

wadded-up paper out of an empty Biro tube, which is basically the same thing. I bet if you got the big garden peas that'd be like the cannonballs of pea-shooting. It's all very wholesome. The rest of the countryside, it's all go hard or go home. But East England is more like go easy, then go home and have a nice cup of tea. That said, they do take it seriously. The champions have laser sights on their pea shooters. It's like *Zero Dark Thirty* starring the Bash Street Kids, and who wouldn't want to see that?

By the time you notice the little red dot, it's too late for you.

BECCLES WORLD THUMB-WRESTLING CHAMPIONSHIPS

This is another thing we used to do for fun in school that adults now take very, very seriously. They even make little wrestling rings for their thumbs to do battle in. I bet some of their thumbs can deadlift a hundred and twenty kilos. Really hench thumbs, saying to the other thumbs, 'Do u even lift bro?'

'1-2-3-4, I declare a thumb wa ... wait, those nails are too fabulous to fight.'

GREAT FINBOROUGH RACE OF THE BOGGMEN

This is where two Suffolk villages race each other on Easter Monday, because more than a hundred years ago, a farmer decided to sack his workers from Great Finborough and hire workers from Haughley in their place. The two groups agreed to race for the job. Mind you, apparently the first lot of workers were sacked for drunkenness. And whoever got to the pub first while carrying the employment contract were the winners. So I'm not sure this would have been any kind of solution for anyone. This is why we now have employment laws, and HR departments. Imagine your livelihood depending on racing through a load of muddy fields. I'd be stuffed. Said it before, will say it again: I'm not built for racing, I'm built for lifting.

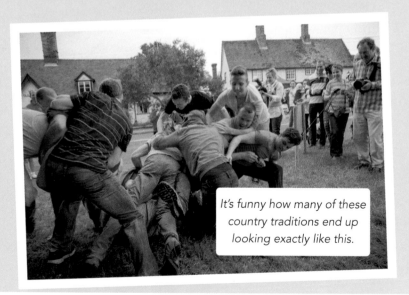

It's funny how many of these country traditions end up looking exactly like this.

LINCOLN CRYING CHRISTMAS

I'm assuming from this that in Lincoln, they spend most of the Christmas holiday in tears? I don't blame them. Three days in and I usually just want to sit down and weep. There's only so much turkey and Monopoly you can stand before you're hankering to get back out on a tractor. Of course, there's always the possibility I may have misunderstood, and this actually refers to a splendid and dignified event dating back at least to the sixteenth century and maybe beyond, in which a group of medieval-style musicians called the Lincoln Waites lead a parade through the city's night-time streets, often pausing to read out a Christmas proclamation, alongside dancing and old-fashioned songs. But if it is, then nobody told me.

'Funny you should mention that,
because that's exactly what it is.'

Chapter Ten

North &
North-East
England

There's something about this region that goes under the radar. Everyone knows about Newcastle, it's true, where they have a tradition called Black Eye Friday, which is the last Friday before Christmas when everybody goes out and gets hammered – in every sense of the word. But that's less an organized event, more of an informal occasion. Again, in every sense of the word 'informal'. You don't have to dress up for it, unless football tops count. But the reputation is unfair, because when you actually meet anyone from Tyneside, they're always some of the most good-natured and friendly folk in the country.

The Likely Lads, twenty-first-century edition.

'Cumbria, Northumberland, I'm gonna make you a star, baby!'

Newcastle is the major city, and the rest of the region doesn't big itself up that much. On your Instagram page – or at least on mine because I follow loads of country accounts – you see Cornwall and Wales and Scotland all the time, and people are always filming TV shows in those places. But aside from the Lake District, which is all very scenic and is probably the geographical equivalent of an influencer, you don't see a lot about that whole area of England just below the border with Scotland. I wonder if they like it that way. Let everybody else get on with shouting about themselves, while they just carry on doing their own thing. But you know what – I'm going to do my best to put them on the map, whether they like it or not.

EGREMONT CRAB FAIR WORLD CURNING CHAMPIONSHIPS

This is a contest to see who can get the most applause while sticking their head through a horse's collar and pulling a face. I'll say this for Cumbria, they're obviously not frightened of the wind changing. Because when I pulled faces as a kid my mum or my nan would always say, 'Don't do that, the wind will change and you'll stay that way forever.' (My mates sometimes ask me if that's what happened, the cheeky sods.) It scared me off doing it, but they're not so easily put off in Cumbria. They've got a whole competition for it, which is especially impressive when you think how windy it is up there. It's the only place in England that has a wind with its own name. That is, its own official name: the Helm wind. Obviously, lots of winds have unofficial names, like 'Aargh, you b*****d!' as

This actually happened and you have to say Her late Majesty was nothing if not a good sport.

they drive freezing rain sideways through your gilet while you're out in some waterlogged field. The Helm wind has been described as 'a horizontal whirlwind', so it changes direction constantly. I bet Cumbria is full of ex-contestants from the gurning competition who look that way all the time now. It's probably a brilliant place to go if you want to impress girls simply by not having your mouth next to your left ear and your eyebrows down by your chin.

The fair itself goes back to 1267, so it's one of the oldest anywhere, which could mean people have been wandering around Cumbria for nearly eight centuries looking like that. That might explain why all those beautiful photos of the Lake District never have anyone in them. Still, it's good to know that gurning has a history stretching far beyond going to raves and taking certain substances that are only supposed to be given to horses – you know, to calm them down after some Newcastle fan has tried to fight them, or whatever. Maybe in the old days Cumbria discovered some kind of mead that had the same effect and they've been at it ever since.

'Forsooth, verily I am 'avin' it!'

ALLENDALE TAR BAR'L

If there's one thing that doing this book has taught me, it's this: we, as a nation, absolutely bloody love setting fire to things. And the further north you go, the more we love it. Maybe it's because we're quite far north in the first place and we get a lot of long, dark nights, so setting fire to things has a lot of bonuses: warmth, light, plus it's one hell of a laugh when you're p*ssed. Doing this book, I think I've found more fire festivals than any other kind. Every two minutes, somebody goes, 'F*** it, let's have another fire festival! Let's jump through a bonfire!' (They actually used to do this until recently in Whalton, in Northumberland, at the Baal Fire, but now they

It was written in the scrolls that this would happen.

just have children crowd around it and dance, which is so much safer.) 'Oh dear, I seem to be on fire myself. Never mind, it's all traditional and good fun!' Another thing I've found out is we love doing stuff with barrels: carrying them, rolling them, throwing them around, and so on. So, I suppose it's inevitable that somebody would combine the two things: get some barrels, then set fire to them.

But they don't stop there. Oh no. That wouldn't be anywhere near dangerous enough. If Cumbria's all about gurning, then Northumberland's all about burning. So, on New Year's Eve in Allendale, forty-five local blokes put on fancy dress and parade through the village carrying whisky barrels filled with burning tar on their heads. Now, if this happened round my way and I was invited to take part, the first thing I'd do is sort out some of the potholes. But perhaps that's not as much of an issue in Allendale, where the roads have probably melted completely by now. You need to have been born in the Allen Valleys to be a guiser, as they call themselves, and many of them have inherited their roles. Just imagine, you're a kid, happy as Larry, doing normal kid stuff, marbles, conkers, PlayStation, and you get this tradition handed down to you out of the blue: 'Son, you need to put this costume on and go carry a barrel full of burning pitch on your noggin', then throw it onto a bonfire.' 'Aw, Dad, do I have to?' 'YES.' 'Can't my sister do it instead?' 'NO GIRLS.' Only one woman has ever been allowed to join in. Her name was Vesta Peart, and she carried a barrel in the 1950s as

How is everything in this photo not already on fire not already on fire?

a reward for making lots of the costumes. Which, if you ask me, is a bit of a funny way to say thank you – 'Here you go, love, your chance to melt your own skin off' – but to each their own. Some of her costumes are still in use now, seventy years later. I've got to say, they do look brilliant, but what did she sew them from, asbestos?

PONTELAND
WHEELBARROW RACE

I thought I'd be really good at this because I had to do it often enough in PE at school, although there was always an argument between me and the other person as to who would be the wheelbarrow. They'd complain if they had to run on their hands and I'd point out that if we swapped over then they'd have to carry both their own weight *and* mine, and that usually shut them up.

I was even more pleased to find out that it involves a literal wheelbarrow, because I am a wheelbarrow ninja. I've been using them since I was a tiny little kid. It's still in teams of two – you

Yeah, it was never like this at school, though, was it?

take turns pushing the other person – and I reckon I could find a partner who's just as good as I am. We can't all handle a wheelbarrow. Believe me, it isn't easy. They tilt, they wobble, they overturn on the slightest bump or change in camber.

We may not be that good at driving a Bugatti Veyron or a Lamborghini Countach with our pinky finger, unlike some

'Who left that *!@%ing pebble there?'

We used a black and white photo because the licence is cheaper.

people I could mention, but put us behind a wheelbarrow and we're Lewis Hamilton crossed with Luke Skywalker. People like us – wheelbarrow aces, that is, although in Northumberland – do this race every New Year's Day outside the Blackbird Inn. (Of course, there's a pub behind it; there always is, God bless 'em.) It goes ahead no matter what the weather, even if it's in four feet of snow. I am totally down for that. One Wheel Good, that's my motto.

SANTON BRIDGE WORLD'S BIGGEST LIAR CONTEST

I hate liars. Just tell me the truth. But what I like about this – apart from the fact that politicians and lawyers aren't allowed to enter on the grounds that they're professionals – is that it's not really lying, more bullsh*tting. You know, spinning tall tales for fun. Stuff about travelling to Scotland in an amphibious wheelie bin, or the way the Cumbrian sugar mines boost jam production. One guy won the contest with a story about how the Lake District and all the local mountains had been nicked from Essex, which is why Essex now looks like it does. Plus, it was once won by a TV celebrity, Sue Perkins, which makes me think I might stand a chance. Trouble is, the most incredible stories I can think of involve the way Jeremy farms, and all of those happen to be true. It's a moot point anyway, as the competition closed down recently – or so they tell us, but I'm not sure if we can believe them, can we?

Sue Perkins, preparing to tell us that she is not in fact Sue Perkins but several owls in a jacket disguised as Sue Perkins.

NORHAM BLESSING OF THE SALMON FISHING

Back when fishing was a big part of life on the River Tweed, people used to gather at the start of the season to see the local vicar bless the nets. They don't fish with nets any more, but they do fish with rods, and now the vicar wades out into the water – in his cassock, which I really hope isn't dry-clean-only – to bless the salmon fishing season itself. After the blessing everybody gets a bit of shortbread and a 'wee dram', which would normally be a bit much for eight in the morning, but I'd say that if you've been sloshing about in a freezing river in early February you've – excuse my language, vicar – bloody well earned it. I'm all for it, anyway, because I love salmon, although my interest in salmon fishing mainly goes as far as watching those wildlife programmes from Alaska where the bears scoop them up as they head up the river. It's how the world works. The natural rotation of life is amazing. Just in case you thought I was going to break into an Elton John song there, I'm really channeling my inner David Attenborough. Obviously, he's a god to me, unlike some other television celebrities I could mention. If they ever need someone to talk over a nature documentary and he's too busy, they can call me and I'll be there faster than a fish leaping up a waterfall.

'...and last year I caught one that was this big.'

OVINGHAM GOOSE FAIR

This is a fairly ordinary and light-hearted kind of village fair, only it features a giant goose, which is the one thing in the entire world more frightening than a normal goose. Geese are made up of approximately twenty per cent fat and eighty per cent pure hatred. The fat is beautiful, and I don't ever feel guilty about eating goose, because they are utterly evil and I'm pretty sure they'd do the same to me given the slightest chance. The only advantage you have when fighting a goose – and be in no doubt, if you see a goose, or rather if a goose sees you, then you will end up either fighting a goose or running away from a goose – is your size, and you can't even count on that because geese have enough rage in them to take down things five times as big. So, if I was standing there enjoying the parade and then I saw an eight-foot goose heading towards me, I'd be out of Northumberland like you-know-what through a you-know-what-else, and back in Oxfordshire before it even reached the spot where I'd been standing.

'Aiieeee! It's heading this way! Run! RUN!'

Chapter Eleven

Yorkshire

A delegation from Yorkshire sets out to protest the insufficient prominence of the region within this volume.

I had to give Yorkshire a chapter to itself. Partly because it's the biggest county in England. At least, North Yorkshire is, and when you add in the other Yorkshire counties the whole area is *yuuge*. But mainly because I'd never hear the end of it if I didn't. Not only from Jeremy, who comes from there and seems to think this is a good advert for the place, but also from every Yorkshire person I ever meet.

I know four things about Yorkshire. Firstly, everyone hates putting their hand in their pocket, and they're proud of it. Anyone else, if you said that, they'd think you were insulting them. But Yorkshire people are all, 'Too right, lad! Eat all, sup all, pay nowt.' I don't even know what that means, but most of the things they say sound just like that. They've almost got their own language, and they say things like, 'Where there's muck

'By 'eck, there goes t'bugger owes me ten pound. Better aim for legs or I'll never see it.'

there's brass', and 'Never do owt for nowt'. Which I think means you can get paid to do dirty jobs, and it's true – that's how I make my living, cleaning up the messes of particular people from Yorkshire – and that you should make sure you always do get paid, which is also true. So, I suppose I've got a lot in common with them, really.

The second thing I know about Yorkshire is, out of all the people who say they're farming 'man's land', which as I mentioned happens a lot, nobody does it more than farmers from Yorkshire: 'You're on boys' land, but we're on man's land.' As usual, I don't know what that means, but I don't even try arguing with them, because the third thing I know about Yorkshire is that there's absolutely no point arguing with somebody from Yorkshire. They'll just say a load of stuff that sounds like ''appen grackle pickle spackle', then walk away looking really pleased with themselves, as if they've just dropped the mic in front of you. The fourth thing I know is a really strange fact, which is that anything stuffed seems to be made up there, whether it's a Paddington Bear or a DFS sofa. Jeremy's especially proud of that last one. Which is funny because when you ask him if he's actually got a DFS sofa he goes very quiet for once. Yeah, that's what I thought.

There's a sofa under there somewhere, but we're not sure if that one's from Yorkshire either.

YORKSHIRE DAY

This is the sort of thing I'm talking about. Does any other county do this? Because if they do, I've not seen it. True, everybody has their own regional pride and their own traditions, and rightly so, but only Yorkshire has decided to have a day named after itself. I think that's what's known as protesting a bit too much. Almost as if they're insecure about something. Each year, they read the Declaration of the Integrity of Yorkshire, which makes me wonder, mate, who are you trying to convince? You know, I'm not going around saying Yorkshire has no integrity – you don't need to tell me. They do this at the four bars of York, which sounds like a decent pub crawl, until you find out it refers to the different city gates. Maybe they go for a p*ss-up when they're finished. I do hope so, because almost every event in this book involves a p*ss-up at some point. That's more or less the whole point of these events: how to make a drink with your mates more interesting. If my friends made me declare the integrity of Oxfordshire at the landmarks of Chipping Norton (which I think would have to include the farm shop and the zoo if you wanted to get up to a total of four), they'd bloody well better buy me a pint afterwards.

Hang on, the what championships?

THE YORKSHIRE PUDDING BOAT RACE

I love a Yorkshire pudding. Tell a lie, I love a dozen Yorkshire puddings. I love all the Yorkshire puddings. I'll leave it for actual pudding and have treacle on it. If you put gravy on it, we're not friends anymore. So, I am very grateful to Yorkshire for inventing it. It's basically an inflated pancake, but most of the great ideas are simple ones. It's a contribution to the world that makes up for Jeremy's ideas about farming. Nearly.

I thought I knew Yorkshire puddings, but that was before I saw the ones they use here in Brawby. I've never met Simon Thackray, but I know he's a genius, a visionary, because he's the one who came up with the idea of a Yorkshire pudding big enough for a kid to paddle down a river in, and then made

drools in Northern

it a reality. Fifty eggs, twenty-five pints of milk, four bags of flour – there you go. Now, if you wouldn't just mind making an extra one for me, and leaving off the coating of yacht varnish that keeps the water out, I'll just sit and eat that while I watch the race, and everybody will be happy. This is absolutely brilliant. I'm going to tell Oscar that his training starts now. If you were a kid and you went into school on Monday and you hadn't done this, but some of the other kids had, you'd be so, so disappointed. This is one of the very best things I've ever heard of. I don't blame Mr Thackray for putting a trademark on it, which to be fair is a very Yorkshire thing to do – 'That's ours, and you're not having it!' – because when you've come up with an idea this good you've got every right to protect it. Wonderful.

FERRET LEGGING

This, apparently, is an ancient pastime that was revived by Yorkshire miners in the 1970s. Because going a mile underground and digging out flammable rocks while breathing air full of dust in a tiny tunnel with an atmosphere that could blow up at any minute isn't nearly difficult or dangerous enough – you really have to challenge yourself in your free time. And what better way to do it than by sticking a ferret down your keks and seeing how long you can keep it there?

'Trust us, mate, it isn't our idea of a good time either.'

While not wearing any underpants, just to make it interesting. No. No no no no no. I know ferrets. I was bitten by one once. On the finger. And that was bad enough. I like ferret racing, that's fun, although I keep well away from the evil little creatures. But ferret legging? This has to be Yorkshire people trying to prove they're harder than everyone else. Give them their due, you do have to be nails, or mad, or both, to do this. You're still not Scotland, though, where if they stick a ferret down their trousers and it bites them on the unmentionables it'll probably lose its teeth.

GAWTHORPE WORLD COAL-CARRYING CHAMPIONSHIPS

This is another legacy of the mining industry, obviously. The story goes that it started as an argument between two coal merchants as to who was in better shape. Nowadays, you'd settle it in the gym, but those were more rugged times. On

Easter Monday, they have separate races for men, women and children, who run a thousand metres to the village maypole carrying a load of coal: fifty-kilo sacks for the men, twenty for the women, ten for the children. As you can imagine, everybody's completely knackered by the end of it. Nice of them to include the kiddies, perhaps as a reminder of the bygone days of child labour. It definitely beats sending them up a chimney.

'And no, that's not me in the purple jumper.'

ECTON BRIDGE GOOSEBERRY SHOW

I'm told this is the oldest surviving gooseberry show in the whole country. Which came as a surprise, I'll admit, because I didn't know there were any surviving gooseberry shows in the country – or even that there ever had been any to survive or not. The world record for heaviest gooseberries has often been set here. The current holder is serial winner Graeme Watson, whose Millennium variety gooseberry in 2019 weighed in at 64.83 grams. Wow. That's a big berry. You'd be proud to have a berry that big.

We pity the fool who messes with those gooseberries.

Keep our traditional fruit and veg competitions clean, we say.

Apparently, the Gooseberry Society have special scales they use to weigh the fruit. I wonder if that's a Yorkshire thing, because Jeremy has some really fancy scales in his kitchen. I'm obsessed with nicking them so I can go and weigh the calibrations on my drill. He hates it. Whenever he can't find the scales he always comes down to the farm and asks me about it, even though Lisa's probably put them in a different cupboard. I bet they need top-notch equipment at the Gooseberry Show. As with any keenly contested competition, they must have dealt with cheaters in their time, injecting water into the gooseberries and the like. Or maybe steroids. If your gooseberry has deltoids and a six-pack, there's obviously something not right.

SLAITHWAITE MOONRAKING FESTIVAL

There aren't many biennial events in this book, but I'm all for inclusivity, so I'm glad we've managed to get this one in. It's held every second February. It's inspired by a local story about some smugglers who had hidden their haul in the pond, and were trying to retrieve it when the authorities happened by. So the smugglers pretended to be idiots who thought the reflection of the moon was a giant cheese, and were trying to rake it out of the pond. Most people think it's a myth, but not me. To my mind, it's one hundred per cent true. You could actually see that happening up in Yorkshire. Because they're canny enough to know that if you want to get away

Frankly, we'd rake a pond for that any day. *drool*

with something clever, you should pretend to be stupid. Also, hiding things in the most obvious place can work really well. Apparently the guy who wrote the *Father Brown* stories once said, the best place to hide a leaf is in a tree.

True or not, they now do a brilliant festival about it, where they make amazing paper lantern figures to carry around the village in a procession, including all sorts of weird fantastical creatures, and a moon that goes off in a shower of fireworks at the end. They put on costumes, beat illuminated drums, carry wooden rakes, play brass band music – the whole thing's completely magic. They obviously really enjoy the idea of putting one over on anyone who tries telling them what to do – you can see where Jeremy gets it from.

We don't know what this has to do with the moon in a pond, and we don't care, because it's amazing.

GREAT KNARESBOROUGH BED RACE

This is just what it sounds like: one person sits on a bed and five other people push them on a three-kilometre course around the town. Which would be fine if it didn't happen to go through a river. I fancied being the one lying on the bed until I found out that bit. Some teams finish the course in around fifteen minutes, which is amazing. I can't run three kilometres in fifteen minutes without pushing a bed. Meanwhile, in a place up on the coast, about an hour away, they've got the Staithes Nightgown Parade. Now, tell me if I'm wrong, but doesn't this sound like an opportunity to meet halfway – in Northallerton, say – and combine the two? Always got my marketing head on, me.

Now, this will give you nightmares about wetting the bed.

MARSDEN CUCKOO FESTIVAL

Is it just me, or does that cuckoo, um, remind you of anything?
I know it's to welcome in the spring, and all that, which means
it probably has origins as a fertility ritual, but, come on. Who
do they think they're fooling?

By 'eck! That's a right big … cuckoo.

Chapter Twelve

Countrywide Customs

Doing this book has shown me that some countryside events are popular wherever you go in Britain. Anything that involves setting fire to stuff, obviously, and anything that involves an excuse for a p*ss-up. We've covered lots of those already. Just about the only thing we don't do is combine the two and set our actual booze on fire, which is something that happens if you go on holiday in Europe. I can only assume drinks are a lot cheaper there than they are here. If anyone comes near my drink with a lighter, let's just say they'll regret it. You might as well try to torch my wallet.

'Hang on, I just paid for those!'

What's nice to see is that, for all our regional differences, there are certain

'What, leave farming?
And give up all this?'

traditions and events that you can find all over the country.
People from different cities hate each other, or pretend to,
but country people tend to get along wherever we go, because
we know our lives are so similar and we like so many of the
same things. We might joke about who's got a harder furrow
to plough – and we mean it literally – but we all know that
farming isn't easy anywhere, and we all love it anyway.

COUNTRY SHOWS

I talked about the particular shows I like the most earlier. But the truth is I like all of them. Mainly, I like that we have them in the first place, and that anywhere you go, there's a county fair, or an agricultural show, so farming communities can get together, show off what they can do, and get a bit drunk on cider. It brings us all together, at the end of the day. And at the start of it, and the middle, too, although the cider part tends to take priority at the end of it. It's how lots of people first fall in love with farming and agriculture, seeing the pride we take in what we do. I also like how each one is so representative of its place: the different breeds of animals you see – like the Oxford Sandy and Black pig, or the Cotswold sheep, round our way – or the different types of farming. In some places, you get hedge-laying competitions. In the flat, arable areas, you get lots of tractors, which is right up my lane, and big sprayers. Farming is so varied, but when it comes down to it, we're basically the same, and you can see both the variety and the common ground at all these shows. That's why I love them.

The Cotswold sheep, where Kaleb got the inspiration for his perm.

MAYPOLE DANCING

This is one of those things that country people everywhere are supposed to do. But I've never, ever done it, and nobody I know has ever done it either. I know it happens all over the place, but I wonder if it's another thing country people mainly do for the tourists, so the tourists can say to one another, 'How quaint, and of course you know it's really an ancient fertility festival and the maypole represents a giant [ahem]'.

Country style…

Which might be true, or might not, and nobody knows, but country people pretend to be all innocent about it. 'Yeah, yeah, we've never heard that before, we're just unsophisticated rustic simpletons – did you bring your money?' The only time I've ever danced around a pole was at a farmers' party and it was a very different kind of pole dancing. It's possible I

Not my usual pole dancing style.

was somewhat drunk. It's possible I was somewhat topless. It's possible it was a telegraph pole. It's possible that I got creosote all over my nipples and they stung for weeks afterwards. But I couldn't confirm any of that.

PRETTIEST BITCH COMPETITION

Now this one really does separate the country from the city. If you say it to anyone in the countryside, they'll know exactly what you mean. Say it in London, you're going to get lynched. Or given a record contract. One or the other. Prettiest Bitch competitions are something that's very countryside. In the city, dogs are pets, but in the country, they're assistants, employees, colleagues – that's not a typing error, although often they're collies as well.

'I'll be with you in a minute, darlings!'

I've been around dogs all my life. My mum's a dog groomer, so I've got to know all breeds of dogs really well from early on. Dogs have been part of my world from a very young age – I don't think there's ever been a time when I wasn't around dogs. I got my own first dog when I was twelve. Usually, kids say, 'Yes, I'll look after it' when they're begging for a dog, then they get one and in two weeks they're bored. But it was different for me. My dog was my working partner, and that's how it's been ever since. There are two types of people in this world. People who walk the dog, and people who get walked by the dog. Or maybe it's that there are two types of dog,

'All right, Mr DeMille, I'm ready for my chew toy.'

and if you put a city dog next to a country dog, you can always tell which is which from their behaviour. And country dogs are like country people – we work all the hours, so we like to go to an event every now and then, and show off. By the way, there's also a 'Most Handsome Dog' category in the same dog shows, just in case anyone thinks we're objectifying bitches.

SCARECROW COMPETITIONS

I think this is one more of those things that city people like to see countryside people do. But I can't deny they are a massive thing. These days we use bird scarers more, but people still love to make scarecrows. I know because every time they have one of these competitions somebody makes a pair that look like me and Jeremy, then they tag me on Instagram. And I'm thinking, f***ing hell, why am *I* getting tagged in this? Leave me out of it! Just do Jeremy, he's the scary one. I'm young and gorge ... well, young, anyway. I'm a young farmer, he's an old wrinkly guy, let *him* scare the birds away.

Not a crow in sight. Job done.

COASTEERING

The first time somebody mentioned this, I thought they said 'castrating' and, honestly, I'm not sure which one's worse. As you'd guess from the name, this is something they do at the coast, and it's become nationally popular ... among lunatics. It's a bit like parkour, that thing when you run up and down buildings in a hooded anorak, or such is my understanding. Only you do it on cliffs and rocks, and you jump into the water. So, no thanks. I'm designed to stay on the floor, and on my feet, and on land – not jumping off things, and especially not into the sea.

Nope nope nope nope nope.

You'd think somebody would have told him there's a guy pushing it in the opposite direction.

BALE-PUSHING

I know a guy who owns a gym. I arm-wrestled him the other day. I didn't even have to try. I'm not saying this to boast. It's just that farm work gives you muscles in places you didn't even know you had muscles, or places. You're going to be constructed differently. Gyms only got invented in the first place because people stopped working on farms. Anyone can choose to lift a weight, but the thing about moving heavy stuff around on a farm is you've got no choice. Want the cows fed? Then you'd better do it, whether you think you can or not. We get so used to it that even when we have some time off, we start doing it for fun. That's how you get bale-pushing contests all over the country. You knock off after a hard day of pushing three hundred-kilo bales around and you meet up with your mates and push some four hundred-kilo bales around. Just because you can.

BALL GAMES

You can go anywhere – literally anywhere – in the whole of Britain and you'll find an ancient ball game where the ball itself is, let's say, incidental at best. Some of them are played on the ground, with the ball at your feet, like football. Some of them, you carry the ball, like rugby. Some of them, you hit or throw the ball around in the air. No surprise that the Scottish have come up with the most dangerous and terrifying version, called shinty, which is like hockey, only you whack a hard ball around the place with sticks at head height, then afterwards you pick up your teeth and go and swig neat whisky out of the winners' trophy. Somebody once told me, Scottishly, that 'It's no a game for weaklings and degenerates', and he was an elderly priest. Scotland. Where even the clergymen are totally hard b******s.

Shinty. Ooft.

Still, shinty is an actual organized sport, with rules and teams and leagues and so on. What most of these ball games have in common – from Hurling the Silver Ball in St Ives in the far South-West to the Kirkwall Ba' in Orkney – is that they're basically an excuse for an all-day riot that only ends when nobody can stand or see any longer. Perhaps the best example is the Atherstone Ball Game in Warwickshire, which has been played every Shrove Tuesday since around the year 1200. Back then, instead of using a ball, they had a bag of gold, which the winners could keep. (The ball tends to be a recent refinement in these games. On the Scottish/English borders, they'd use the head of a Sassenach – an Englishman to you and me – or a Scotsman, depending on which side the game took place on.) Nowadays in Atherstone they've got a specially made ball the size of a small car, and one rule: you can only play in one particular street, which is boarded up for the occasion. In 2023, it was reported that the game descended into violence, with punches thrown and shopfronts damaged. How this was any different to the previous 822 games, I have no idea.

We are shocked – shocked – to learn that such a game could descend into violence.

PANCAKE RACES

I love pancakes, and I love anything to do with pancakes. First, because pancakes are delicious, obviously. But also because Pancake Day should really be called Thank A Farmer Day. I mean, what goes into pancakes? Milk. Flour. Eggs.

'We tried just racing the pancakes against each other, but it was less than trhilling.'

That's three different types of farming right there: dairy, arable, poultry. And if they're managed right, they're all types of farming that help each other – the chicken farmers can provide fertilizers, the dairy farmers can help with the grass and the soil, and so on. Really, there's nothing more British than pancakes. No matter what age you are, you can run a pancake race, or flip a pancake – which I'm quite good at – or just sit down and eat a pancake.

Pancake traditions go back a long way. In Toddington, in Bedfordshire, they've got Conger Hill Witch Listening, when local children go and lie down on the grass to try and hear the witch frying her pancakes inside the hill. That's really sweet, but when I think about it, I remember hearing about the witch hunts we used to have and all the people – all the women,

'If you listen very carefully, children, you can hear the echoes of the shared past that unites us all no matter where we come from, and that's the beauty of tradition.'

especially – who lost their lives. Just put to horrible ends, drowned or burned or hanged for no reason. It reminds me that traditions and customs don't always represent something good. But traditions and customs can change, and over time they do, and they should. They're living things, not something that has to be set in stone. They adapt. A lot of the mummers and Morris dancers that used to black up their faces don't do that anymore, and that's quite right. Some people complain about it, but I think the opposite: what it shows is that we can keep the good bits, which tell us something positive about who we are and where we come from, and lose the bits that don't belong in our own age. This whole book has been a joy to me because I can see how that happens. So, thanks for sharing it all with me.

ACKNOWLEDGEMENTS

First, I want to thank Taya, without you none of this would ever have been possible. Thanks also to Jeremy, we have learnt so many difficult skills from each other on this journey. What a blast it's been so far! Thanks also to the team at Fresh Partners – to Debbie, Ellis, Danielle and Toni, thank you for keeping me on the right track. And big thank you to the team at Plank PR – to Lou and Cecile. And a huge thank you to the whole team at Quercus who have looked after me so well. Thanks also to David Bennun for all your hard work.

Picture Credits

Chapter One: 7 iStock / Getty Images / Raylipscombe; Chris Terry; **8** Shutterstock / 1000 Words; **9** Shutterstock / Kwadrat; **10** iStock / Getty Images / BrianAJackson; **11** Shutterstock / ComposedPix; **12** Shutterstock / SofikoS; **13** Shutterstock / H. Tuller; **14** (top) Alamy / Allstar Picture Library Ltd, (bottom) Alamy / John James; **15** (top) Shutterstock / Nick Biemans, (bottom) Shutterstock / WilleeCole Photography; **16** Shutterstock / Paul Wishart; **17** (top) Shutterstock / mountaintreks, (bottom) Shutterstock / bogdanhoda; **18** Alamy / Haydn Denman; **19** Alamy / Thousand Word Media; **20-21** Alamy / arovingeye / Stockimo; **21** Shutterstock / Khamidulin Sergey.

Chapter Two: 23 iStock / Getty Images / acidgrey, **23** Chris Terry; **24** Alamy / PA Images; **25** (top) iStock / Getty Images / shironosov, (bottom) iStock / Getty Images / SeventyFour; **26** iStock / Getty Images / AlexAndrews; **27** Shutterstock / Hurst Photo; **28** Shutterstock / ANGHI; **29** (top) Alamy / Terry Applin, (bottom) Alamy / Jim Holden; **30** Alamy / Nick Turner; **31** Alamy / Guy Corbishley; **32** Alamy / robertharding; **33** India Rose Creative; **34** Alamy / Aggie Trunglebuckets Image Emporium; **34** India Rose Creative; **35** Shutterstock / alyxfuzz; **36** Alamy / Finnbarr Webster Editorial; **37** Shutterstock / Richard Austin.

Chapter Three: 39 iStock / Getty Images / Peter Fleming, **39** iStock / Getty Images / Halfpoint; **40** Shutterstock / seeshooteatrepeat; **41** iStock / Getty Images / Sasiistock; **42** (top) iStock / Getty Images / Lorado, (bottom) Mark Modra; **43** Alamy / Guy Corbishley; **44** Trionium; **44** Trionium; **46** Alamy / PA Images; **47** Alamy / Gareth Fuller; **48** Alamy / MARTIN DALTON; **49** Alamy / Cinematic Collection; **50** Alamy / Emma Wood; **51** Alamy / Emma Wood; **53** Dave Mariott.

Chapter Four: 55 iStock / Getty Images / Philippe Paternolli, **55** Chris Terry; **56** iStock / Getty Images / Lukas Bischoff; **57** Alamy / Stuart Kelly; **57** iStock / Getty Images / Wendy Love; **58** iStock / Getty Images / bahadir-yeniceri; **59** iStock / Getty Images / FatManPhotoUK; **60** iStock / Getty Images / baiajaku; **61** (top) Peter Barnett, **61** (bottom) Peter Barnett; **62** (top) Alamy / Mark Lewis, (bottom) Alamy / Chronicle; **63** Alamy / Jeff Morgan 04; **64** Alamy / Haydn Denman; **65** Alamy / Clarissa Debenham; **66** Alamy / Steve White; **67** Peter Barnett; **68** Green Dragon Activities Ltd; **69** Alamy / Paul Smith / Featureflash Film Archive; **71** Alamy / Richard Naude

Chapter Five: 73 iStock / Getty Images / R Clare, **73** iStock / Getty Images / Mark Dean, **73** Unsplash / Greg Wilson, **73** Steve McDonough; **74** Alamy / Abbus Archive Images; **75** iStock / Getty Images / Rena-Marie; **76** Averil Shepherd / Calendar Customs; **77** Getty Images / Stanley Bielecki Movie Collection; **78** Shutterstock / Joshua Tudi; **79** Alamy / Deborah Vernon; **80** Alamy / Moviestore Collection Ltd; **81** Alamy / PA Images; **82** Alamy / Jacob King; **83** Alamy / Richard Bradley; **84** Averil Shepherd / Calendar Customs; **86** Alamy / WENN; **87** Clova Perez-Corral; **88** Alamy / PA Images; **89** Alamy / Jeff Morgan 03; **90** Alamy / Wayne Neal; **92** Alamy / Richard Bradley; **93** Caroline Robinson / Artist Jemma Gowland.

Chapter Six: 95 Alamy / Pictorial Press Ltd, Chris Terry; **96** iStock / Getty Images / MediaProduction; **97** iStock / Getty Images / duncan1890; **98** iStock / Getty Images / travellinglight; **99** (top) Pretty Green Photography, (bottom) Shutterstock / Azukari Photography; **100** iStock / Getty Images / Torsakarin; **103** Alamy / Phil Taylor; **104** Alamy / Phil Taylor; **105** Shutterstock / Victoria OM; **105** Shutterstock / Victoria OM, **105** Chris Terry; **106** Alamy / Peter Byrne; **107** Shutterstock / Real Sports Photos; **108** Jonathan White; **110** iStock / Getty Images / dusanpetkovic; **113** Averil Shepherd / Calendar Customs.

Chapter Seven: 115 Alamy / Matt Limb OBE, iStock / Getty Images clubfoot, iStock / Getty Images / chrisbrignell, iStock / Getty Images / GlobalP; **116** iStock / Getty Images / daverhead; **117** iStock / Getty Images / duncan1890; **120** Alamy / Dave Donaldson; **121** Alamy / Lefteris Pitarakis; **122** Alamy / Doug Houghton; **123** Alamy / Doug Houghton; **124** iStock / Getty Images / InfotronTof; **125** Alamy / Ross Gilmore; **126** Shutterstock / Angus Blackburn; **128** Alamy / sleuth110 **128** Shutterstock / Michael Mcgurk; **131** Alamy / WENN.

Chapter Eight: 133 Chris Terry; **135** Shutterstock / Sem / Universal Images Group; **136** Alamy / Findlay; **137** iStock / Getty Images / 5./15 WEST; **138** Shutterstock / BluIz60; **139** iStock / Getty Images / whitemay; **140** Shutterstock / Alexander_P; **141** Alamy / Allstar Picture Library Ltd; **142** Alamy / schubbel; **143** Shutterstock / Bits And Splits; **145** Alamy / Andrew Milligan; **146** Shutterstock / Tina Norris **148** iStock / Getty Images / skhoward; **149** Shutterstock / Ian Melvin.

Chapter Nine: 151 iStock / Getty Images / sharply_done, Chris Terry **152** iStock / Getty Images / Mary Michelle Emery; **153** iStock / Getty Images / ian_parker; **154** Alamy / Homer Sykes; **156** Cereals; **157** Shutterstock / Guy Erwood; **158** Shutterstock / NEOS1AM; **159** Alamy / Emma Wood; **161** Ross Parish / Traditional Customs and Ceremonies **163** Averil Shepherd / Calendar Customs; **164** Shutterstock / Geoffrey Robinson; **166** Alamy / Jeff Spicer; **167** Alamy / Paul Gapper; **169** Ross Parish / Traditional Customs and Ceremonies.

Chapter Ten: 171 iStock / Getty Images / John Longley; **171** iStock / Getty Images / panic_attack; **172** Shutterstock / Hazel Plater; **173** Shutterstock / Axel Alvarez; **174** Alamy / Arthur Edwards; **175** Bounce By The Ounce / SS Movies / YouTube; **176** iStock / Getty Images / colaboy; **178** Neil Denham; **179** iStock / Getty Images / GlobalStock; **180** iStock / Getty Images / Jan-Schneckenhaus; **181** Alamy / Mirrorpix ;**182** Shutterstock / Featureflash Photo Agency; **184** Chris Strickland **185** Averil Shepherd / Calendar Customs.

Chapter Eleven: 187 iStock / Getty Images / miracsaglam, iStock / Getty Images / heibaihu, Shutterstock /HappyTime19, Chris Terry; **188** Alamy / All Star Picture Library Limited; **189** iStock / Getty Images / shank_ali; **190** Getty Images / Chris Jackson; **191** Alamy / capt.digby **192** iStock / Getty Images / from_my_point_of_view; **193** Alamy Amazing Action Photography, iStock / Getty Images / JohnGollop Chris Terry; **194** iStock / Getty Images / Harald Schmidt; **196** Alamy / 2ebill; **198** Alamy / Owen Humphreys, Shutterstock / Ivan_Nikulin **200** Shutterstock / Mushakesa; **201** Chris Chinnock; **202** Alamy / Danny Lawson; **203** Alamy / David Preston.

Chapter Twelve: 205 Alamy/ Nick Maslen, Chris Terry; **206** Shutterstock / Africa Studio; **207** iStock / Getty Images / Diane-Kay **209** iStock / Getty Images / Sonya Kate Wilson; **210** iStock / Getty Images / Oscarhill; **212** Shutterstock / Javier Brosch; **213** Alamy / Andrew Fox; **214** North Wales Live / Ian Cooper; **215** Shutterstock / Gai Johnson; **216** Alamy / Findlay; **218** Shutterstock / JASPERIMAGE; **219** Alamy / Homer Sykes; **220** Alamy / Richard Slater; **221** Shutterstock / Lopolo.

224